Artificial Intelligence and Digital Forensics

Artificial intelligence (AI) has revolutionized several sectors, with digital forensics being one that has been notably affected by its rapid advancement. AI brings previously unheard-of powers to this field, helping authorities examine large datasets, identify developments, and uncover digital evidence that is essential to cracking cybercrimes. Despite these promising developments, using AI in digital forensics presents problems. The complexity and dynamic nature of cyberattacks are a significant challenge, demanding the ongoing adaptation of AI models to new attack strategies. This changing environment makes it difficult to create reliable and future-proof solutions. This book explores the advancements, applications, challenges, and solutions that AI brings to the realm of digital forensics.

Artificial Intelligence and Digital Forensics: Advancements, Applications, Challenges, and Solutions includes the latest applications and examples with real data, making the book meaningful for readers. It is written in very simple language, with each technology having its dedicated chapter that explains how it works and provides an example of a real-world application. Key points and a summary are provided at the end of each chapter to enable the readers to quickly review the major concepts as it presents a practical understanding, so the readers can be better equipped to handle AI-based tools with ease and efficiency. The book provides unconditional support to those who are making decisions by using massive data from their organization and applying the findings to real-world current business scenarios addressing the challenges associated with integrating AI into digital forensics and offering practical solutions. It also offers insights into the ethical use of AI technologies and guidance on navigating legal requirements and ensuring responsible and compliant practices.

This book caters to industry professionals from diverse backgrounds, including engineering, data science, computer science, law enforcement, and legal experts, fostering a holistic understanding of the subject along with providing practical insights and real-world examples.

Artificial Intelligence and Digital Forensics

Advancements, Applications, Challenges, and Solutions

Edited by Naveen Kumar Chaudhary, Archana Patel, Abinash Mishra, and Chothmal Kumawat

CRC Press
Taylor & Francis Group
Boca Raton London New York

CRC Press is an imprint of the
Taylor & Francis Group, an informa business

Designed cover image: Shutterstock

First edition published 2026
by CRC Press
2385 NW Executive Center Drive, Suite 320, Boca Raton FL 33431

and by CRC Press
4 Park Square, Milton Park, Abingdon, Oxon, OX14 4RN

CRC Press is an imprint of Taylor & Francis Group, LLC

ISBN: 978-1-032-84468-8 (hbk)
ISBN: 978-1-032-84471-8 (pbk)
ISBN: 978-1-003-51282-0 (ebk)

DOI: 10.1201/9781003512820

Typeset in Times
by Apex CoVantage, LLC

Contents

Preface xv
About the Editors xvii
List of Contributors xix

Chapter 1 Uses of Artificial Intelligence for Digital Forensics 1

Chethana R.M.

1.1 Introduction to Digital Forensics 1
1.2 Challenges in Traditional Digital Forensics 2
1.3 Role of Artificial Intelligence in Digital Forensics 2
1.4 Applications of Artificial Intelligence in Digital Forensics 3
1.5 Benefits of Artificial Intelligence in Digital Forensics 3
1.6 Ethical Considerations in Artificial Intelligence-Enabled Digital Forensics 4
1.7 Future Trends in Artificial Intelligence and Digital Forensics 4
1.8 Tools Used for Artificial Intelligence Digital Forensics 5
1.9 Software Used for Artificial Intelligence in Digital Forensics 6
1.10 Conclusion 7
References 8

Chapter 2 Challenges in Integrating Artificial Intelligence into Digital Forensics 9

Arunima Negi, Arpita Parihar, Abinash Mishra, and Ambrish Mishar

2.1 Introduction 9
2.1.1 Overview of Digital Forensics 9
2.1.2 Role of Artificial Intelligence in Digital Forensics 10
2.1.3 Importance of Addressing Integration Challenges 11
2.2 Key Challenges in Artificial Intelligence Integration 12
2.2.1 Data Quality and Availability 13
2.2.2 Model Interoperability 14
2.2.3 Scalability Issues 16
2.3 Technical Limitations 18
2.3.1 Lack of Standardized Tools and Frameworks 18
2.3.2 Integration with Existing Digital Forensic Tools 18
2.3.3 Dependence on Computational Resources 19
2.4 Ethical and Legal Challenges 20
2.4.1 Privacy Concerns 20
2.4.2 Bias and Discrimination in Artificial Intelligence Models 21
2.4.3 Compliance with Legal Standards 22

2.5 Security Risks in Artificial Intelligence-Based Forensics 24
2.5.1 Vulnerabilities to Adversarial Attacks 24
2.5.2 Risk of Artificial Intelligence Misuse by Cybercriminals 25
2.6 Human Factors 25
2.6.1 Skill Gap among Digital Forensics Professionals 25
2.6.2 Resistance to Adopting Artificial Intelligence Tools 26
2.6.3 Balancing Automation and Human Oversight 26
2.7 Case Studies and Examples 27
2.7.1 Successful Artificial Intelligence Implementations in Digital Forensics 27
2.7.2 Failures and Lessons Learned 28
2.8 Recommendations and Best Practices 28
2.8.1 Strategies for Enhancing Data Quality 28
2.8.2 Frameworks for Ethical Artificial Intelligence Use in Forensics 28
2.8.3 Training and Upskilling Forensic Experts 30
2.9 Conclusion 31
Bibliography 31

Chapter 3 Towards Open Standards for Systemic Complexity in Digital Forensics 33

Paola Di Maio

3.1 Ubiquity in Digital Forensics 33
3.2 Systemic Complexity 34
3.3 Digital Forensics and Artificial Intelligence 35
3.4 Open Challenges 37
3.5 Reasoning 41
3.6 Towards Open Standards 42
3.7 Model Cards in Digital Forensics 44
3.8 Future Work 46
References 46

Chapter 4 Identifying Patterns through Integration of Natural Language Processing and Semantic Web Technologies 49

Tatiana Avdeenko and Artem Golubnichiy

4.1 Introduction 49
4.2 Fundamentals of Natural Language Processing 52
4.2.1 Morphological Analysis 52
4.2.2 Syntactic Analysis 53
4.2.3 Semantic Analysis 54

4.3 Semantic Web and Ontologies in Pattern Identification Tasks..... 57
4.4 Machine Learning Methods for Pattern Identification 59
4.4.1 Classification Models 59
4.4.2 Clustering Models 60
4.4.3 Time Series Analysis 60
4.4.4 Examples of Using Machine Learning Methods to Identify Patterns 60
4.5 Integration of Pattern Recognition with the Semantic Web and Data Fusion from Heterogeneous Sources 61
4.5.1 Use of Ontologies 61
4.5.2 Data Integration from Various Sources 62
4.6 Algorithm for Using Natural Language Processing Libraries and Semantic Web Technologies for Pattern Identification 62
4.7 Discussion 63
4.7.1 Interpretation of Results 63
4.7.2 Current Challenges and Limitations in Text Analysis Using Natural Language Processing and Semantic Web Technologies 63
4.8 Conclusion 65
References 66
Appendix A: Textual Data for Analysis 68
Appendix B: Example Code for Algorithm 70

Chapter 5 Futuristic Approaches in Conversational AI for the Cloud Platform 72

Panem Charanarur, Srinivasa Rao Gundu, Aditya Mohan Srivastava, H.K. Pratihari, Naveen Kumar Chaudhary, and Laishram Hemanta Singh

5.1 Introduction 72
5.2 Conversational Features of Humans 72
5.3 Morality in Artificial Intelligence-Based Conversation 73
5.4 Advantages of Artificial Intelligence-Based Conversation 74
5.5 Components of Artificial Intelligence-Based Conversation 74
5.6 How Conversational Artificial Intelligence Actually Functions? 74
5.7 Achieving Conversational Artificial Intelligence 75
5.8 Incorporation of Conversational Artificial Intelligence 75
5.9 Conversation Design and Artificial Intelligence Enhancing the User Interaction 75
5.10 Conversational Artificial Intelligence and Chatbot Differences 76
5.11 Difficulties in Conversational Artificial Intelligence 76

5.12 Advanced Conversational Artificial Intelligence 76
5.13 Understanding the Role of Artificial Intelligence and Machine Learning in the Future of Computing 77
5.14 Purpose of the Study 77
5.15 Objectives of This Study 78
5.16 Why Artificial Intelligence and Machine Learning Are Crucial for Future Computing? 78
5.17 A Prospective Model for Future Computing Systems 78
5.18 Model within Its Own Autonomous Controls 78
5.19 Next-Generation Computing with Explainable AI (XAI) 79
5.20 Possible Dangers Inherent in Artificial Intelligence-Based Integrated Computing 80
5.21 Concussions and Future Perspectives 80
5.22 Statement of Funding 80
5.23 Conflict of Interest Statement 80
References 80

Chapter 6 Harnessing the Power of Artificial Intelligence to Enhance Digital Forensics 82

Raunak Sood

6.1 Introduction 82
6.2 Usage of Artificial Intelligence in Acquisition of Evidence 83
6.2.1 Case Analysis 85
6.3 Safeguarding Privacy in Artificial Intelligence-Based Digital Forensics 86
6.3.1 The United States 86
6.3.2 The European Union 87
6.3.3 India 88
6.3.4 Comparative Analysis 89
6.4 Digital Forensics and Privacy Issues 89
6.5 Artificial Intelligence in Digital Forensics 91
6.6 Way Forward 91
6.6.1 Case Analysis of Cellebrite Guardian [22] for Privacy Management and Evidence Management 91
6.6.2 Workings of the Cellebrite Guardian 91
6.6.3 Case Analysis of Veritone Redact Where Artificial Intelligence Is Being Used to Protect Privacy When Used in Digital Forensics 92
6.7 Conclusion 93
References 93

Chapter 7 Analyzing Vast Datasets with Machine Learning 96

Vineet Bhandari, Ganti Sri Rama Krishna, Chothmal Kumawat, and Narayan C. Debnath

7.1 Introduction 96
7.1.1 Understanding Vast Datasets in Digital Forensics...... 96
7.1.2 Characteristics of Vast Datasets.................................. 97
7.1.3 Sources of Vast Datasets... 97
7.1.4 Challenges in Handling Vast Datasets........................ 98
7.1.5 Role of Machine Learning in Managing Vast Datasets 98
7.2 Machine Learning Fundamentals for Data Analysis 98
7.2.1 Overview of Machine Learning.................................. 99
7.2.2 Supervised Learning... 99
7.2.3 Unsupervised Learning... 99
7.2.4 Reinforcement Learning (RL)................................. 100
7.2.5 Core Concepts in Machine Learning for Data Analysis.. 100
7.2.6 Basic Types of Machine Learning Algorithms......... 101
7.2.7 Decision Trees 103
7.3 Data Visualization in Machine Learning 104
7.3.1 Data Visualization in the Pre-Modeling Phase......... 104
7.3.2 Explainability and Interpretability through Visualization.. 109
7.3.3 Challenges and Future Directions in Data Visualization.. 111
7.4 Types of Analysis Using Machine Learning.......................... 112
7.4.1 Classification Analysis 112
7.4.2 Clustering Analysis 113
7.4.3 Anomaly Detection.. 113
7.4.4 Regression Analysis 113
7.4.5 Dimensionality Reduction....................................... 114
7.4.6 Time Series Analysis 114
7.4.7 Hybrid Approaches.. 114
7.5 Applications of Machine Learning in Digital Forensics 115
7.5.1 Data Analysis and Evidence Extraction 115
7.5.2 Malware Analysis.. 115
7.5.3 Incident Response and Intrusion Detection............. 116
7.5.4 Timeline Reconstruction 116
7.5.5 Imaging and Video Forensics 116
7.5.6 Textual Evidence Analysis 117
7.5.7 Behavioral Profiling 117
7.5.8 Swarm Intelligence in Evidence Analysis................ 118

7.5.9 Predictive Analysis for Crime Prevention 118
7.5.10 Digital Evidence Validation 118
7.5.11 Automation of Repetitive Tasks 119
7.5.12 Managing Big Data in Forensic Investigations 119
7.5.13 Inspection of Files and Formats 120
7.6 Challenges in Analyzing Vast Datasets with ML 120
7.6.1 Data Quality and Preprocessing Challenges 121
7.6.2 Computational Challenges 121
7.6.3 Feature Engineering and Dimensionality Challenges 121
7.6.4 Model Interpretability Challenges 122
7.6.5 Scalability and Infrastructure Challenges................. 122
7.6.6 Ethical and Legal Challenges 122
7.6.7 Visualization Challenges 123
7.7 Conclusion 123
Bibliography 124

Chapter 8 Artificial Intelligence and Forensic Intelligence: Analyzing Social Behavior through Historical Data... 126

Mahyuddin K. M. Nasution

8.1 Introduction 126
8.2 Background: Behavior, Treatment, and Sources 126
8.2.1 Individual Behavior 127
8.2.2 Organizational Behavior 128
8.2.3 Government Behavior 129
8.2.4 Social Behavior 129
8.3 Natural Law and Behavior 130
8.3.1 Water, Words, and Deeds 130
8.3.2 Customs and Culture 131
8.4 Methodology 131
8.4.1 Documents and Data 132
8.4.2 Social Behavioral Forensics 132
8.5 Role of Artificial Intelligence 133
8.6 Forensic Intelligence.. 136
8.6.1 Corruption Personal and Social Behavior 136
8.6.2 Community with Behavior 137
8.6.3 Analysis 140
8.7 Conclusion 141
Bibliography 141

Chapter 9 Identifying IOCs from Compromised Android Devices 149

Prince Bhuva and Archana Patel

9.1 Introduction 149
9.1.1 Malware 149
9.1.2 Overview of Android 150
9.1.3 Malware Analysis Methods 150
9.2 Literature Review 151
9.3 Methodology 153
9.3.1 Testing Environment 153
9.3.2 Malware Analysis Methodology 153
9.4 Malware Analysis Using Santoku: Implementation 154
9.4.1 Santoku 154
9.4.2 Installing and Configuring Genymotion 155
9.4.3 Static Analysis with VirusTotal 155
9.4.4 Dynamic Analysis with Cuckoo Sandbox 156
9.4.5 Analysis with VirusTotal 156
9.5 Results 157
9.6 Conclusion 157
References 158

Chapter 10 Technology Tools for Legal System: Understanding the Gaps and Enhancing the Capacity of the Stakeholders 159

Kapil Saini, Manmeet Dhillon, and Ravi Saini

10.1 Overview of Our Legal System 159
10.2 Understanding the Stakeholders' Needs and Challenges 160
10.2.1 Judges 160
10.2.2 Lawyers 160
10.2.3 Litigants and Parties 161
10.2.4 Legal Support Staff 161
10.2.5 Law Enforcement Personnel 162
10.3 Imperative of Technology Integration in the Indian Legal System 162
10.3.1 AI-Powered Systems 162
10.3.2 Data Analytics Platforms 163
10.3.3 Document Automation Software and Case Management Systems 164
10.3.4 Virtual Courtroom Technology, E-Filing Systems, and Client Portals 164
10.3.5 Online Dispute Resolution (ODR) Systems 164

10.4 Modernization of the Indian Legal System ... 165
10.4.1 eCourts Project ... 165
10.4.2 Legal Information Systems ... 166
10.4.3 Case Management Software ... 166
10.4.4 Virtual Legal Assistant Applications ... 166
10.4.5 Legal Cloud Storage or Legal Document Management Systems ... 166
10.5 Challenges in Adoption and Implementation of Technology Tools ... 166
10.5.1 Digital Divide and Technological Illiteracy ... 167
10.5.2 Reliability and Accuracy ... 167
10.5.3 Digital Security and Privacy Concerns ... 167
10.5.4 Big Data Management and E-Discovery ... 167
10.5.5 Training and Capacity Building ... 167
10.5.6 Disinformation and Misinformation ... 168
10.5.7 Cybercrimes and Jurisdictional Challenges ... 168
10.5.8 Lack of Standardization ... 168
10.5.9 Artificial Intelligence and Automation ... 168
10.5.10 Regulatory Barriers ... 168
10.6 Successful Implementation Strategies ... 169
10.6.1 Developing Clear Objectives and Requirements ... 169
10.6.2 Ensuring Data Privacy and Security ... 169
10.6.3 Addressing Ethical and Legal Implications ... 169
10.6.4 Pilot Study and Continuous Evaluation ... 170
10.6.5 Training Programs for Legal Professionals ... 170
10.6.6 Collaborations and Other Initiatives by Institutions ... 170
10.7 Summary ... 171
References ... 171

Chapter 11 Future of Digital Forensics: Implications of Transparency and Accountability in Artificial Intelligence ... 173

Raunak Sood, Sarthak Sood, and Ranjit Karmahapatra

11.1 Introduction ... 173
11.2 Transparency in Artificial Intelligence-Driven Forensics ... 174
11.3 A Global Perspective to Understand the Legal Challenges in Digital Forensics Connected with Artificial Intelligence ... 175
11.4 Analyzing the Impact of Lack of Transparency in AI-Driven Forensics: A Case Study of Image Analyzer ... 178
11.5 Accountability in Artificial Intelligence-Driven Digital Forensics ... 179
11.6 Intersection of Ethics and Law in AI-Enabled Digital Forensics ... 182

11.7 Policy Recommendations and Way Forward 184
11.8 Scope for Future Work: AI-Based Data Translation 185
11.9 Conclusion 185
References 185

Chapter 12 Legal and Ethical Imperatives for Transparency 191

Devanshu Sharma and Dr. Navtika Singh Nautiyal

12.1 Introduction 191
12.1.1 Importance of Transparency in Fostering Trust........ 191
12.1.2 Relevance of Legal and Ethical Considerations in Artificial Intelligence Systems 192
12.1.3 Transparency's Role in Digital Forensics and Decision-Making Systems....................................... 192
12.2 Legal Frameworks for Artificial Intelligence Transparency.. 192
12.2.1 Global Laws and Regulations Promoting Transparency in Artificial Intelligence 193
12.2.2 Other International Artificial Intelligence Laws on Transparency 194
12.2.3 Case Studies: Consequences of Non-Transparency in Artificial Intelligence 194
12.2.4 Comparative Analysis of Legal Approaches 195
12.3 Ethical Foundations of Transparency 195
12.3.1 Ethical Theories for Transparency........................... 196
12.3.2 Duty to Explain the Results of AI Analysis............. 196
12.3.3 Human Autonomy Evolution.................................. 196
12.3.4 Improving Organizational Accountability 197
12.3.5 Balancing Ethical Considerations with Conflicting Interests Review 197
12.3.6 Risk of Misuse... 197
12.3.7 Balancing Ethical Trade-Offs 198
12.4 Challenges to Achieving Transparency 198
12.4.1 Technical Challenges.. 198
12.4.2 Resistances from Organizations towards Opening Up of Their Secret Algorithms 199
12.4.3 Lack of Standardization and Global Agreement on Transparency Norms... 199
12.4.4 To Overcome the Challenges.................................. 200
12.5 Implications of Lack of Transparency 200
12.5.1 Risks of Algorithmic Bias, Unfair Decision-Making, and Discrimination 201
12.5.2 Loss of Trust among Users and Stakeholders 201
12.5.3 Legal and Ethical Consequences of the Black Box.. 202
12.5.4 Implication of the Black Box 202

12.6 Proposed Guidelines and Standards for Transparency 202
12.6.1 Analysis of International Rules 203
12.6.2 Recommendations for Implementation of Transparency in Artificial Intelligence Systems 203
12.6.3 Government-Industry-Academia Collaboration 204
12.6.4 Towards a Unified Standard 205
12.7 Future Trends in AI Transparency 205
12.7.1 Emerging Technologies to Enhance Transparency 205
12.7.2 Hypothesized Future Legal and Ethical Frameworks 206
12.7.3 Issues while Formulating Consensus-Based Norms 206
12.7.4 Area for Improvement 206
12.8 Conclusion 207
12.8.1 Meaning of Transparency Legal and Ethical Issues 207
12.8.2 Barriers to Transparency End 208
12.8.3 Key Recommendations 208
12.8.4 A Call to Involvement 208
References 209

Index 211

Preface

The book *Artificial Intelligence and Digital Forensics: Advancements, Applications, Challenges, and Solution* highlights advanced machine learning and deep learning techniques and its real-time applications in digital forensics. In addition, it outlines the various use cases, challenges, and machine learning-driven solutions for the digital forensic community. Last but not the least, it improves the forensics practices and enables readers to build a reliable and robust automated decision-making system in the current digital forensic era.

It presents the fundamental ideas of integrating artificial intelligence for digital forensics to resolve contemporary issues of forensics and practical use of AI in evidence acquisition. It is important to find the patterns in digital evidence for their comprehensive context for data analysis and evidence correlation. For the same, semantic web and natural language processing (NLP) technologies are discussed along with forensics intelligence for examining the pattern of activity of criminals and their network by using the historical data. It also includes principal concepts and methods along with applications of ML pertinent to big data management, focusing on approaches of mandates complexity in big mass of datasets, including feature engineering, dimensionality reduction, and removing noise. The book also covers the role of technological tools, implications of transparency and accountability in artificial intelligence, and its legal and ethical imperatives for enhancing the working of the legal system.

This book provides a roadmap to explore the advances, obstacles, solutions, and implications of AI in digital forensics as well as discover the optimal use of AI in digital forensics to provide highest benefit to the user under one comprehensive source and title.

About the Editors

Naveen Kumar Chaudhary has been a professor of cybersecurity and dean at the National Forensic Sciences University, Gandhinagar, Gujarat, India, since 2019. He is also a courtesy research professor at the Knight Foundation School of Computing and Information Science, Florida International University, Miami, USA. He holds a bachelor of technology degree in information technology and telecommunication engineering, a master of engineering degree in digital communication, and a master's in business administration. He earned his Ph.D. in engineering and advanced certifications in cyber and network security. Dr. Chaudhary has extensive experience spanning more than 26 years in engineering education, research, and government and has worked extensively on the formulation and implementation of policies and techno-legal frameworks in the domain of cybersecurity, e-governance, and forensic science. He is actively involved in promoting start-up ecosystems for developing low-cost indigenous cyber-forensic solutions. He is editor-in-chief of the NFSU *Journal of Cyber Security and Digital Forensics*, an IEEE senior member, an honorary life member of the Soft Computing Research Society, and a life member of IETE. He is the recipient of a letter of appreciation for his contribution towards the cause of literacy from Brent St. Denis, MP, Algoma, Canada, in 1994, and the Dubai SEWA award for his contribution to cybersecurity education in 2022.

Archana Patel is an assistant professor at the National Forensic Sciences University, Delhi, India. She has worked as a full-time faculty member at the School of Computing and Information Technology, Eastern International University, Binh Duong Province, Vietnam. She has completed her postdoc at Freie Universität Berlin, Berlin, Germany. She published a patent entitled "Method and System for Creating Ontology of Knowledge Units In A Computing Environment" in November 2019. Dr. Patel received a doctor of philosophy (Ph.D.) in computer applications and a master's degree both from the National Institute of Technology (NIT) Kurukshetra, India. She has qualified GATE and UGC-NET/JRF exams. Dr. Patel has also contributed in a research project funded by the Defence Research and Development Organization (DRDO) for two years. She is an author or co-author of more than 40 publications in numerous referred journals and conference proceedings and has been awarded the best paper award (four times) at international conferences. She has served as a reviewer in various reputed journals and conferences. Dr. Patel has received various awards for presentation of research work at various international conferences and teaching and research institutions. She has edited 12 books, served as a guest editor in many well-reputed journals, and delivered various keynote talks. Her research interests are artificial intelligence, ontological engineering, semantic web, big data, expert systems, and knowledge warehouse.

Abinash Mishra received his Ph.D. in 2022 from the Department of Computer Applications, National Institute of Technology, Tiruchirappalli, Tamil Nadu, India. He joined the National Forensic Sciences University, Delhi, India, as an assistant professor in 2022 and served as an assistant professor at the Indian Institute of Information Technology Nagpur before joining the NFSU. His research interest lies in applying various artificial intelligence approaches to extract useful patterns from different facets of societal challenges of high demand with economic benefit. The main aim of his work is to develop machine learning and deep learning models for various domains like biometrics, forensic science, banking, telecom, and healthcare. He has published various articles in reputed journals and conference proceedings.

Chothmal Kumawat received a B.Tech. degree in electronics and communication engineering from Government Engineering College, Ajmer, India, in 2012, and an M.Tech. in communication systems and Ph.D. degrees from the Indian Institute of Technology Roorkee, Roorkee, India, in 2014 and 2023, respectively. From July 2014 to May 2015, he was an assistant professor at the Department of Electronics and Communication Engineering, Malviya National Institute of Technology, Jaipur, India. Now, he has been working as an assistant professor with the School of Cyber Security and Digital Forensics at National Forensic Sciences University, Delhi Campus, New Delhi, India, since November 2022. His current research focuses on multimedia forensics.

Contributors

Tatiana Avdeenko
Novosibirsk State Technical University
Russia

Vineet Bhandari
School of Cyber Security and Digital Forensics, National Forensic Sciences University, Delhi Campus
New Delhi, India

Prince Bhuva
National Forensic Sciences University, Delhi Campus
India

Panem Charanarur
Department of Cyber Security and Digital Forensics, National Forensic Sciences University Tripura Campus
Tripura, India

Naveen Kumar Chaudhary
School of Cyber Security and Digital Forensics, National Forensic Sciences University Gandhinagar
India

Narayan C. Debnath
Eastern International University
Vietnam

Manmeet Dhillon
Geeta Global Law School, Geeta University
Panipat, Haryana, India

Artem Golubnichiy
Novosibirsk State Technical University
Russia

Srinivasa Rao Gundu
Department of Digital Forensics, School of Sciences, Malla Reddy University
Hyderabad, Telangana, India

Ranjit Karmahapatra
Defence Cyber Agency (DCyA, HQ IDS, Ministry of Defence)

Ganti Sri Rama Krishna
School of Cyber Security and Digital Forensics, National Forensic Sciences University, Delhi Campus
New Delhi, India

Chothmal Kumawat
School of Cyber Security and Digital Forensics, National Forensic Sciences University, Delhi Campus
New Delhi, India

Paola Di Maio
Center for Systems, Knowledge Representation and Neuroscience
USA

Abinash Mishra
School of Cyber Security and Digital Forensics, National Forensic Sciences University, Delhi Campus
New Delhi, India

Ambrish Mishar
Member of International Society for Development and Sustainability (ISDS)
Japan

Mahyuddin K. M. Nasution
Computer Science Department, Universitas Sumatera Utara
Indonesia

Navtika Singh Nautiyal
Post Doctorate from MICHR
Italy

Arunima Negi
School of Cyber Security and Digital Forensics, National Forensic Sciences University, Delhi Campus
New Delhi, India

Arpita Parihar
School of Cyber Security and Digital Forensics, National Forensic Sciences University, Delhi Campus
New Delhi, India

Archana Patel
National Forensic Sciences University, Delhi Campus
India

H. K. Pratihari
Department of Cyber Security and Digital Forensics, National Forensic Sciences University Tripura Campus
Tripura, India

Chethana R.M.
AIML Department, CMR University,
Bangalore Karnataka, India

Kapil Saini
Assistant Professor, School of Computer Science Engineering and Applications, Geeta University
Panipat, Haryana, India

Ravi Saini
Lead Scientist, Deloitte US
USA

Devanshu Sharma
National Forensic Sciences University
India

Laishram Hemanta Singh
Department of Cyber Security and Digital Forensics, National Forensic Sciences University Tripura Campus
Tripura, India

Raunak Sood
School of Criminal Law and Military Law, Rashtriya Raksha University (An Institution of National Importance), Ministry of Home Affairs
Rashtriya Raksha University (An Institution of National Importance), Ministry of Home Affairs
Gujarat, India
Bennett University (Times of India Group)
Greater Noida, India

Sarthak Sood
Advocate, Solan District Court
Solan, Himachal Pradesh, India

Aditya Mohan Srivastava
Department of Cyber Security and Digital Forensics, National Forensic Sciences University Tripura Campus
Tripura, India

1 Uses of Artificial Intelligence for Digital Forensics

Chethana R.M.

1.1 INTRODUCTION TO DIGITAL FORENSICS

Digital forensics is the collection, examination, and presentation of electronic evidence. As information becomes quickly digitized and electronic devices increase, there is also a need for digital forensics within cybercrime, fraud, etc. [1]. With the rapid growth of digital data and the wide use of electronic devices, digital forensics is more crucial than ever before. This procedure uses the application of computer investigations and analysis techniques to gather a body of evidence.

The first step is to collect digital evidence from computers, mobile phones, tablets, servers, and other devices. The type of data captured in these artifacts include documents, emails, images, videos, browsing history, and system logs. Digital forensic experts use special tools and expert techniques to collect such evidence while keeping it intact. They manage its chain of custody to ensure its admissibility in a court of law [1]. After they are collected, the obtained evidence is analyzed to reveal useful information and prove the facts of the case. The analysis verb examines files.

Structures, studying metadata, cracking codes, and piecing together digital activity to recreate timelines and sequences of events. Additionally, digital forensic experts rely on more sophisticated approaches such as data carving, steganalysis, and network forensics, which help recover erased or hidden information that might be critical to the investigation.

Ultimately, it is all about very clearly and compellingly communicating the findings that are to stand up to legal scrutiny. This is done through documentation, organization of found evidence, and production of comprehensive reports with the methods used and conclusions reached. Digital forensic specialists are also sometimes required to testify in court as expert witnesses to help decrypt complicated technological terms and articulate authoritative interpretations of the evidence presented. The incidences of cybercrime, fraud, and other illegal practices that are now multiplying in the digital arena make digital forensics a more integral part of law enforcement and trial practice. Digital forensics relies on technology and expertise to retrieve digital evidence that may be used to ensure proper justice [1], hold perpetrators accountable, and protect the integrity of digital environments.

DOI: 10.1201/9781003512820-1

1.2 CHALLENGES IN TRADITIONAL DIGITAL FORENSICS

Traditional digital forensics largely relies on manual analysis, which is too time-consuming and prone to error. Besides, the mammoth volumes and varieties of digital information generated every day are a challenge to investigators.

- Limitations of manual analysis: In manual computer evidence analysis, very specialized skills and subject knowledge are involved, thus making it a costly and prone-to-error technique. Investigators must sift through mountains of data [2], which becomes impractical and overwhelming in complex cases.
- Volume and variety of digital data: The addition of digital data, mostly in electronic media posts and emails, increases difficulties in traditional forensic analysis approaches [2]. Different formats and different methods of stored data make extracting and analysis processes complicated.

1.3 ROLE OF ARTIFICIAL INTELLIGENCE IN DIGITAL FORENSICS

Traditional digital forensics involves meticulous examination of digital devices and data to uncover evidence of cybercrime or unauthorized activities. However, this process occurs often. It will become complex with data explosion because cyber threats in today's world are evolving with time with a need to respond promptly [1]. AI technologies offer a great hope to address these challenges. One of the major advantages of AI in the context of digital forensics is that such technologies allow automation at most stages of processing. Machine learning algorithms can be programmed to analyze and identify patterns and anomalies in digital data so that they can scan enormous amounts of information very quickly [2]. This automation expedites the process of investigation under the hands of forensic analysts as they can concentrate their efforts on more critical tasks that require human interventions [2].

AI-driven analytics can make forensic investigations even more effective and accurate by including not only log data and network traffic but also file metadata. AI learns to identify intricate signals, indicating malicious activity that may not be noticed with the naked eye. The systems can continue to learn from a stream of new data and changing new threats, and the results are directly drawn from that activity. Even that can aid in the extraction and interpretation of digital evidence. For example, NLP algorithms decompose digital text evidence existing in electronic mail evidence, chat logs, or documents so that it may be possible to identify the content worthy of being used in an investigation [2]. Computer vision algorithms break images and videos to extract information such as details about objects of interest.

The software of AI technologies could substantially beautify the performance and effectiveness of virtual forensic teams in the fight against cybercrime. The fundamental concept of automation of recurring duties in addition to superior talents of analytics in AI permits investigators to study more complex instances a lot faster and more effectively regardless of ergonomics. However, such use must be ethical and responsible, especially about questions of information privacy, transparency, and duty. In short, AI integration goes much into revamping the virtual forensic space while heightening efforts directed in the direction of cybersecurity [2] in this more and more digital global.

- Automated data analysis: AI-based tools can also analyze digital evidence automatically, which means that investigators can process big sets of data much faster with higher accuracy. Using algorithms, machine learning platforms discover patterns, trends, and anomalies in data, which helps identify suspicious sporting activities.
- Pattern recognition and anomaly detection: AI algorithms do well in pattern detection as well as knowing when things have gone wrong. AI algorithms can make it much easier to determine anomalous events resembling cyberattacks, data breaches, or unauthorized access, which would lead to an early response.

1.4 APPLICATIONS OF ARTIFICIAL INTELLIGENCE IN DIGITAL FORENSICS

Integrating AI generation in virtual forensics has added numerous applications at some point of great domain names, along with cybersecurity, law enforcement, and employer investigations.

- Malware detection and analysis: AI algorithms can recognize and classify malware threat patterns and behaviors. In this regard, machine learning models help forensic investigators identify new and newly emerging mistypes, thus developing lines of defense in advance [2].
- Image and video analysis: AI-based multimedia-proof evaluation gear can take a look at video or picture surveillance, virtual pics, and more. It is feasible for this equipment to routinely tag, classify, or beautify visible content, supporting investigators in pinpointing the precise facts they are looking for.
- Network traffic analysis: AI Network Forensic Tools: AI-based network forensic tools monitor and analyze network traffic to detect suspicious activities and potential breaches in security. Thus, investigators can apply machine learning algorithms on network data to identify anomalous patterns that reflect malicious behavior, such as intrusion attempts or data exfiltration [2].

1.5 BENEFITS OF ARTIFICIAL INTELLIGENCE IN DIGITAL FORENSICS

Using AI for virtual forensics will offer a wonderful selection of more suitable efficiency, scalability, and accuracy in investigative tactics.

- Increased efficiency and accuracy: AI-powered forensic gear has the potential to automate many time-consuming techniques, along with extraction, analysis, and correlation of statistics, therefore allowing people to pay attention to higher-level analysis and decision-making. In addition, AI algorithms are manipulated to reduce human errors and biases, thus generating greater, constant, and reliable outcomes.

- Handling large volumes of data: It is through scalable AI technologies that thousands of digital pieces of evidence can be processed and analyzed efficiently by forensic investigators. AI systems allow parallel processing and distributed computing and, therefore, can manage different types of data sources and formats, which ensures complete investigations.
- Real-time monitoring and response: AI-based forensic solutions can provide real-time monitoring and alerting capabilities, thus enabling organizations to identify proactive ways to react to cyber threats. Continuous analysis of digital activity and network traffic performed by AI systems can identify suspicious behavior patterns and trigger a mechanism for the automatic responses relevant to such identification.

1.6 ETHICAL CONSIDERATIONS IN ARTIFICIAL INTELLIGENCE-ENABLED DIGITAL FORENSICS

While AI gives terrific capability to decorate digital forensic practices, its adoption will increase crucial moral worries regarding privateness, bias, and duty.

- Privacy concerns: The application of AI algorithms in digital forensics poses a challenge concerning privacy in the collection, storage, and analysis of personal data. The investigators, therefore, have to uphold the ethical and legal standards of data protection and privacy even as they are transparent and seek consent [2].
- Bias and Fairness in Algorithms: AI algorithms are prone to biases in either training data or model-developing processes. Therefore, forensic investigators have to critically appraise the fairness and accuracy of AI-driven decisions and search for potential biases that would impact the interpretation or digital evidence use [3].
- Transparency and accountability: Forensic assurance and the transparency/opacity of AI algorithms: To assure transparency, investigators must ensure the use of best practices and documentation standards that are transparent to help ensure the suitability and reproducibility of AI-driven analyses.

1.7 FUTURE TRENDS IN ARTIFICIAL INTELLIGENCE AND DIGITAL FORENSICS

AI technologies are being used to detect fraudulent activity. New developments to help stop identity fraud and other activities are in the works.

- Advancements in machine learning algorithms: Ongoing research and development in machine learning algorithms, including deep learning and reinforcement learning, will drive innovation in methodologies and tools for digital forensics [2]. Progress of this nature will enable a much more detailed analysis of digital evidence with increased proficiency in detection and attribution.

- Integration of AI with blockchain technology: Combining AI and blockchain technology has the potential to enrich the integrity and authenticity of digital evidence. Blockchain's decentralized and tamper-resistant traits can be used together with AI-enabled forensic structures to set up verifiability and immutability in investigation findings, strengthening its fidelity in court trials.
- Enhancements in forensic tools and software: Forensic tools and software will continuously evolve to apply artificial intelligence to make interfaces more intuitive, efficient, and accessible. Platforms integrating AI-based analytics, visualization, and collaboration functionalities will help forensic investigators optimize workflow for better productivity [2].

1.8 TOOLS USED FOR ARTIFICIAL INTELLIGENCE DIGITAL FORENSICS

AI digital forensics, in turn, is an articulated term from AI-driven digital forensics – the application of artificial intelligence technologies in the investigation and analysis of various forms of digital evidence relevant to numerous cybercrime contexts. In AI digital forensics, a vast gamut of tools and methodologies are applied to make investigative practices more effective and efficient. Some of the principal tools frequently used are listed below:

1. Machine Learning Algorithms: Machine learning algorithms help in the autocompletion of procedures used to investigate digital evidence, which makes them the base of AI digital forensics. They help learn from immeasurable amounts of datasets, thus training them for patterns, anomalies, and malicious behaviors that would alert the investigator towards possible threats or illegal actions.
2. Natural Language Processing (NLP) Tools: NLP gear is commonly useful in content-primarily based digital evidence that consists of chat logs, emails, social media posts, and documents. This helps investigators extract relevant data, examine sentiment, and find linguistic styles that could imply the behavior and motives of the criminal.
3. Image and Video Analysis Software: Multidimensional multimedia files will be analyzed for tampering, manipulation, or hidden information through the use of AI-powered image and video analysis software. It will automatically determine if images have been forged, videos altered, or other digital manipulation has taken place and will assist in the verification of digital evidence.
4. Deep Learning Frameworks: They also permit the development of advanced synthetic intelligence models using libraries such as TensorFlow and PyTorch, thereby facilitating functions that can be achieved in complicated manners applied in virtual forensics, for instance, photo identification, object detection, and voice evaluation. The frameworks assist in the creation of bespoke solutions tailored to specific investigative desires.

5. Forensic Analysis Tools: Augmented with artificial intelligence, forensic analysis tools have empowered investigators to extract, parse, and analyze data from as many diverse digital sources as computers, smartphones, and even cloud storage systems. The tools use artificial intelligence algorithms to make concealed information visible, reconstruct digital actions, and establish timelines of occurrences.
6. Cyber Threat Intelligence Platforms: Cyber threat intelligence platforms facilitate the gathering, analysis, and integration of threat data through AI technologies from a variety of sources, such as OSINT, dark web monitoring, and security feeds. Platforms of this type help investigators identify relevant threats, understand attributes, attack vectors, and attribute cyber incidents to a specific threat actor or group.
7. Blockchain Analysis Tools: In instances concerning cryptocurrencies or primarily blockchain-based transactions, AI-pushed blockchain evaluation gear is used to trace and examine financial transactions, identify illicit sports, and map out transaction networks. These gears assist investigators in uncovering evidence of cash laundering, fraud, or other monetary crimes performed during the blockchain era [4].
8. Behavioral Analytics Solutions: Artificial intelligence algorithms are used in behavioral analytics solutions to examine the patterns of user activity and distinguish those that are abnormal, as they may be indications of illegal or unauthorized actions. With such a system, you can display and monitor insider threats, unauthorized access attempts, and other risky activities targeting your networks and systems [4].
9. Case Management Systems: AI-powered case management structures provide investigators with equipment to effectively create, prioritize, and control cases in digital forensics. It employs AI algorithms to automate ordinary responsibilities and streamline workflows and resources in collaboration among institutional individuals involved with the research method.
10. Data Visualization Platforms: AI-powered fact visualization gear turns complex records into smooth-to-apprehend pics. These gears assist researchers in spotting patterns and seeing tendencies and percentages of what they have found out using charts, graphs, and maps you could have interacted with.

With these AI-enhanced tools and technologies being used in digital forensics, investigators will be able to better combat cybercrime, faster incident response, and enhancement of cybersecurity defense. One should not forget that humans need the right amount of training and know-how to use these tools to their best advantage, thus reducing mistakes or wrong interpretations.

1.9 SOFTWARE USED FOR ARTIFICIAL INTELLIGENCE IN DIGITAL FORENSICS

Artificial intelligence is increasingly being included in digital forensic techniques as a good way to enhance overall performance and effectiveness in the evaluation and interpretation of large amounts of information. Some of the sizable software programs that rely on AI in digital forensics include X–1.

1. Ways Forensics: X-Ways Forensics incorporates AI-based algorithms for obligations inclusive of document carving, picture popularity, and metadata analysis. It enables forensic investigators to correctly process massive volumes of virtual evidence and identify applicable information for investigations [5].
2. Autopsy: Autopsy is an open-source digital forensic platform built using AI techniques for automatic keyword search, image recognition, and pattern-based analysis. It helps forensic examiners recognize digital evidence within various sources of data collected from hard drives, mobile devices, cloud storage, and many other places [5].
3. Cellebrite UFED: Cellebrite uses the Universal Forensic Extraction Device (UFED), which is AI-driven, to support the extraction and analysis of data from smartphones, tablets, and other mobile devices. Using machine learning algorithms, the UFED identifies and categorizes each type of digital evidence [5].
4. EnCase Forensic: EnCase Forensic software by OpenText integrates AI-powered functionalities, particularly for data triage, evidence prioritization, and timeline to obvious potential routine tasks and provides leads to obvious potential fraud, misuse, or other suspicious behaviors for efficient processing of digital investigations.
5. Volatility: Volatility is probably one of the most commonly used memory forensics frameworks that include AI analysis techniques [6]; it allows forensic analysts to identify malware, rootkits, and other malicious artifacts, hidden in volatile memory, using machine learning algorithms and pattern recognition.
6. Paliscope: Paliscope is an online digital investigative tool, using some AI-driven capabilities for social media analytics and image recognition with data visualization. It helps gather evidence from Internet-based sources, including social media, websites, and cloud services, across the digital spectrum, which was crucial in building the case [6].
7. TensorFlow Forensics: TensorFlow is an open-source machine learning framework developed by Google. Research in digital forensics has been adapted using TensorFlow, and TensorFlow forensics provides tools and algorithms concerning image and text classification to detect anomalies [5].

These software application packages provide a reason behind how the generation of AI has extra high-quality virtual forensic techniques in allowing forensic investigators to correctly analyze virtual evidence to discover hidden styles and pick out applicable information for criminal investigations and prison complaints.

1.10 CONCLUSION

Artificial intelligence in digital forensics is a departure from conventional investigation methodologies and is the dawn of a new era within this field. AI will enable professionals in forensic science to use tools that are unmatched by the present ones for scrutinizing, understanding, and showcasing digital evidence. Investigators can overcome long-time hurdles by using AI while amplifying the speed and precision of analysis and hardening their capacity to foresee and counter emergent cyber threats.

However, moral considerations and openness are critical concerns in the software of AI in the realm of digital forensics. This could avoid the miscarriage of AI capabilities while, at the same time, stopping unwarranted results. The inclusion of AI might improve the capacity of research via the speedy processing of humongous record volumes and patterns that are hardly decipherable with the aid of the human mind. This no longer speeds up investigations but also renders findings tons deeper and more accurate, providing an extra expansive view of the virtual footprint and the feasible malware.

However, with the increasing role AI plays in digital forensics, privacy, bias, and accountability become significant issues. Ethical frameworks need to be carefully created and followed without deviation so that the AI algorithms function benignly, transparently, and even with defined accountability. The integrity and credibility of forensic analysis require strict ethical adherence and mechanisms that ensure fairness, accuracy, and due process.

Ultimately, appropriate funding in AI for digital forensics is essential to ensure greater efficacy and robustness against evolving cyber threats. However , such capability needs to be balanced with a similarly strong remedy for ethics and transparency, as this would, in addition, promote public acceptance, trust, and self-assurance of the integrity of forensic practices in the digital age.

REFERENCES

[1]. Jarrett, Aaron, & Choo, Kim-Kwang Raymond (2021). The impact of automation and artificial intelligence on digital forensics. *Wires Forensic Science*, Citation: 06.

[2]. Hall, Stuart W., Sakzad, Amin, & Choo, Kim-Kwang Raymond (2021). Explainable artificial intelligence for digital forensics. *Wires Forensic Science*, Citation: 07.

[3]. Solanke, A. Abiodun (2022). Explainable digital forensics AI: Towards mitigating distrust in AI-based digital forensics analysis using interpretable models. *Forensic Science International: Digital Investigation*, Volume 42, Supplement, 301403.

[4]. Dunsin, Dipo, Ghanem, C., Mohamed, Ouazzane, Karim, & Vassilev, Vassil (2024). A comprehensive analysis of the role of artificial intelligence and machine learning in modern digital forensics and incident response. *Forensic Science International: Digital Investigation*, Volume 48, 301675.

[5]. Bhawna, & Mahajan, Shilpa (2024). Artificial intelligence perspective on digital forensics. In *Applying Artificial Intelligence in Cybersecurity Analytics and Cyber Threat Detection*. Wiley Online Library, Chapter 03.

[6]. Vasilaras, Alexandros, Papadoudis, Nikolaos, & Rizomiliotis, Panagiotis (2024). Artificial intelligence in mobile forensics: A survey of current status, a use case analysis, and AI alignment objectives. *Forensic Science International: Digital Investigation*, Volume 49, 301737.

[7]. Mishra, Anand Kumar, Hemamalini, V., & Tyagi, Amit (2024). Digital forensics with emerging technologies: Vision and research potential for the future. In *Conversational Artificial Intelligence*. Wiley Online Library, Chapter 37.

2 Challenges in Integrating Artificial Intelligence into Digital Forensics

Arunima Negi[1], *Arpita Parihar*[1], *Abinash Mishra*[1], *and Ambrish Mishar*[2]
[1]School of Cyber Security and Digital Forensics, National Forensic Sciences University, Delhi Campus, New Delhi, India.
[2]Member of International Society for Development and Sustainability (ISDS), Japan.

2.1 INTRODUCTION

2.1.1 Overview of Digital Forensics

Digital forensics encompasses the methodical process for identifying, protecting, analyzing, and presenting digital evidence in a manner that is legally admissible. This domain represents a unique genre in forensic science, where an investigator analyzes cybercrimes and a myriad of other illegal acts involving digital equipment. During the past few decades, digital forensics have experienced tremendous change, driven by technological advancement and the increasing prevalence of digital equipment.

This includes the widespread use of mobile devices, cloud storage solutions, and the Internet of Things (IoT) in daily life, which introduces new forms of digital evidence to investigation procedures. In digital forensics, potential evidence may include any electronic data such as email, SMS, web-browsing history, digital files, and encrypted communication. Basically, the main aim of digital forensics is to identify and note the digital evidence to make up a coherent account about an incident; very often, such a scenario either accepts or denies the claim presented in the criminal investigation or legal dispute.

The process of digital forensic investigation typically involves the following steps:

- Identification: It is the process involved in determining the location of digital evidence and methodologies for its acquisition. This includes the collection of evidence from electronic devices, network logs, cloud infrastructures, and other data storage systems.
- Preservation: This includes ensuring that the data does not get modified in the process of extraction. It is done by the creation of forensic images,

DOI: 10.1201/9781003512820-2

which are the exact replicas of digital devices to sustain the integrity of the original information.

- Analysis: Testing the retained data to extract meaningful information. This could include file restoration, a study of communication patterns, harmful activity detection, or even finding hidden information.
- Presentation: Documenting findings in an understandable and concise report. Results need to be presented in a way that can be understood by lawyers and their solicitors, judges, or jurors so that the forensic technique can stand up to examination in a judicial environment.

The importance of digital forensics has increased alongside cybercrimes. Cyberattacks become more sophisticated and digital frauds widespread with each passing day; thus, the need for this realm of proficient professionals who could work with the complex applications of digital evidence has never been higher.

2.1.2 Role of Artificial Intelligence in Digital Forensics

Artificial intelligence is transforming the face of so many things in this world, including digital forensics. AI has transformed how evidence is gathered, analyzed, and ascertained through digital forensic investigation. The forensic process, to date, has been very time-consuming when dealing with large sets of data. AI augments the forensic expert's capabilities because of its ability to learn patterns, adapt to new information, and automate complex processes.

One of the most important applications of AI in the world of digital forensics is its ability to process large volumes of data at incredible speed. AI algorithms are able to identify patterns in data that could elude a human examiner. This is particularly useful when there is a high volume of data – such as in corporate settings, espionage, or significant cybercrime. In such instances, AI-based tools can scan emails and documents, metadata, and net traffic much faster than the time it would take a human investigator to carry out the same task. For example, in fraud or financial crime investigations, AI may be applied to identify patterns of transaction data that are unusual or secret relationships involving seemingly unrelated actors. AI can also be applied in cybersecurity to discover malware or to identify whether there has been unauthorized access into a network.

Tools involving artificial intelligence, such as the work of machine learning algorithms, can automatically classify and categorize data, identify possible signs of criminal activity, and draw attention to suspicious activities for further investigation. Artificial intelligence tools, in the case of social media, can scan text, images, and video content quickly for possible evidence of harassment, threats, or any other illegal activities involved.

In addition, AI can be used to validate digital evidence by scanning for any evidence of tampering or alteration. It is capable of verifying if the digital files are untampered with and unaltered, which is an integral part of evidence integrity, particularly in court presentation. Moreover, AI-based applications have widely been accepted in many fields, such as facial recognition and biometric analysis, and help the sleuth identify suspects or victims involved in any criminal case [1, 2].

2.1.3 Importance of Addressing Integration Challenges

It is well known that artificial intelligence benefits the digital forensic environment greatly, but less clear is how AI technologies can be integrated into existing forensic procedures. Overcoming these challenges will help ensure AI can be integrated into forensic inquiries without sacrificing the integrity of the process or the dependability of the results. With AI in digital forensics, many hurdles arise from technological complexity, data quality, and the need to satisfy legal and ethical requirements. For instance, the "black-box" nature of so many frameworks for AI tools makes it difficult to understand how exactly they arrive at a specific determination. Lack of transparency in these makes it challenging for verification purposes to be assured that results put forth by AI will be considered admissible as evidence in courts and can be adequately articulated to a judge or jury.

Ultimately, overcoming the challenges accompanying the induction of AI into digital forensic activities will be pivotal for maximizing potential benefits and minimizing accompanying risks. Only through the resolution of these challenges can digital forensics evolve to answer increasing needs in the digital environment.

- **Making It More Accurate and Reliable**: An important motivation for dealing with the integration challenges is to make AI more accurate and trustworthy. Digital forensic activities include investigations where there is a high stakes factor in such a way that the evidence should meet stringent legal requirements. Issues of data quality, model interoperability, and algorithmic bias could be used in challenging the admissibility of the evidence obtained from AI systems. In this respect, dealing with these challenges is crucial so that the forensic conclusions are both scientifically valid and legally admissible, for instance, using approaches for explainable AI (XAI), which will facilitate transparency to make it easier for analysts to understand the verification produced by AI.
- **Operational Efficiency and Scalability**: AI can process enormous volumes of digital evidence much faster than more conventional methods can. Challenges to this potential, though, include integration issues like computational resource constraints and the absence of any standard frameworks. It is here that scientists would be able to dissolve these problems to unlock the efficiency and scalability that AI has to offer. For instance, optimizing algorithms that work efficiently in terms of resources and allow them to integrate seamlessly with existing forensic tools would make it possible to process large volumes much faster.
- **Efficiency in Operations and Scalability**: AI can process incredibly large amounts of digital evidence much faster than traditional methods can. Still, there's an integration challenge in terms of lower computational resources and the non-existence of standardized frameworks that hinder such potential. Such obstacles might unlock the efficiency and scalability benefits of AI. For example, algorithms may be optimized to make the best use of resources. Further integration with existing forensic tools would also provide faster processing of large datasets; it would make investigations more time-efficient.

- **Ethical and Legal Compliance**: Ethical and legal issues remain an area of concern, such as violations of privacy and algorithmic bias to date. This may cause the failure of appropriate and ethical use of AI technologies. Eventually, this would signify a loss of confidence on the part of citizens as well as erosion of forensic procedures. Stakeholders might build ethical rules proactively through compliance with different legal requirements to mitigate risks and have increased confidence in AI-based integrated forensic systems. For example, privacy-preserving methods such as federated learning can allow sensitive information to be protected while yet permitting the execution of an all-encompassing forensic analysis.
- **Mitigating Security Risks**: AI systems are vulnerable to adversarial attacks and malicious use by cybercriminals. Such risks could compromise investigations through the introduction of false evidence or manipulation of results. Thus, achieving robustness in adversarial defenses as well as in secure implementation practices becomes essential for preserving the integrity of forensic investigation. More importantly, devising robust AI models that can identify and counter adversarial inputs assures the reliability of findings even in hostile scenarios.
- **Bridging the Skill Gap**: Such skills are required on both sides – technical and qualitative. There is an imbalance in existing skills among forensic experts and technology gurus that holds implementation back. The gap can be addressed only by training and education in such a manner that professionals become ready to apply AI tools effectively and ethically. Integrative education programs in AI, data science, and digital forensics potentially develop the workforce for these needs.
- **Assisting Future Developments**: Solving integration problems will build on future advancements in artificial intelligence and digital forensics. In any trajectory of AI development, overcoming current problems ensures that the domain remains adaptable and open to new applications of new technologies. For example, advancements in natural language processing and computer vision can provide new sources for the analysis of text and image evidence, encouraging creativity in forensic practice.

Through a deep investigation of these integration challenges, the digital forensic discipline has the potential to draw on all that AI can provide to standards of ethics, legality, and operations. The approach not only enriches the present application of AI but also readies the field for highly advanced technology in the future.

2.2 KEY CHALLENGES IN ARTIFICIAL INTELLIGENCE INTEGRATION

Data plays a crucial role in integrating AI into digital forensics. AI lives in data, and in this case, the quality and availability of data are crucial for forensic investigations. This is because, with low-quality and unreliable data, AI models would fail to perform correctly, resulting in false conclusions and backbreaking legal miscarriages. Two of the biggest data quality and availability challenges are inconsistent and noisy data and the availability of large, diverse datasets.

2.2.1 Data Quality and Availability

Data is at the core of the infusion of AI in digital forensics. By the nature of AI, data is deeply ingrained; hence, the quality and availability of data are crucial in forensic analysis. Poor quality and dubious data will hamper AI performance, driving the AI model towards wrong conclusions that may likely compromise legal processes as well. Among the major issues relating to data quality and availability are the presence of dirty and noisy data, as well as the need for large, diverse datasets.

2.2.1.1 Inconsistent and Noisy Data

Inconsistent and noisy data is one of the primary challenges that forensic investigators face when integrating AI into digital forensics. Noisy data refers to information that contains errors, inaccuracies, or irrelevant content, making it difficult for AI models to identify meaningful patterns. In digital forensics, noisy data may stem from a variety of sources, including corrupted files, incomplete records, or irrelevant digital artifacts. For example, in the case of a cybercrime investigation, investigators may encounter logs or data from network traffic that are filled with irrelevant or repetitive entries, complicating the analysis process.

Indeed, AI models, especially those based on machine learning, are data-dependent. The more accurate, clean, and well-labeled data they are given, the better their predictions will be. Nevertheless, noisy or inconsistent datasets hamper the training procedure and may cause inaccurate models, as well as some misinterpretations in data. For example, in spam email fraud, AI may mark legitimate emails as spam if the training dataset contains irrelevant or redundant information. Similarly, if a dataset used to train an artificial intelligence model in identifying malware contains extraneous files or incorrectly labeled instances, the AI is likely to fail in the identification of malware.

[3] Perhaps one of the notable examples of this problem is when one tries to analyze data concerning the trafficking of children. The files are usually inconsistent and dispersed, which explains why they become hard to analyze. Furthermore, the non-existence or corruption of some factors in the data makes it impossible for AI tools to reconstruct the data in a reliable manner, thus making investigators sift through irrelevant information to find credible ones.

Addressing noisy data typically requires advanced data-cleaning methodologies, which encompass the elimination of extraneous data, the rectification of inaccuracies, and the normalization of formats. Furthermore, artificial intelligence systems need to be trained to identify and eliminate noise while prioritizing relevant patterns. Nonetheless, this procedure of cleansing and preparing data can be labor-intensive and may necessitate an advanced degree of proficiency in data science and artificial intelligence. Furthermore, the artificial intelligence system must be adept at managing noisy datasets, necessitating resilient algorithms capable of differentiating between noise and significant information.

2.2.1.2 Access to Large, Diverse Datasets

[4] The largest barrier to implementing artificial intelligence in digital forensics is the acquisition of large and diversified datasets. Large quantities of data are often needed to train AI adequately. Most investigations in digital forensics involve a variety of

different types of data, such as images and videos, textual documents, system logs, and communication streams on network appliances. For AI to be useful in digital forensics, the data must form a solid representation of broad real-world scenarios so that the AI model can generalize and detect patterns characteristic of criminal activity.

However, acquiring different types of datasets is very challenging in various ways. First, data is often broken or proprietary, sometimes even encrypted, which leaves digital forensics workers struggling to collect sufficient data to facilitate the training of artificial intelligence. Other problems are those related to the issue of privacy and the law, where the required datasets needed for the same are filled with sensitive or personal information that cannot be freely shared. For instance, in case of cybercrime investigations, datasets involving data logs or sensitive financial information may not be available because of privacy laws such as GDPR or other regulatory requirements.

The second challenge while obtaining large datasets is that the types of data collected may be either scarce or hard to collect. For example, there may be insufficient bases of attack vector-related data to train AI models to identify new malware or hacking techniques. Another kind of incident where sufficient data may not be easily available is in online harassment and cyberbullying, where it becomes tough to gather sufficient data to train AI models on the language to recognize abusive language across different social media platforms.

One real-world example is the fact that facial recognition AI fails to pick out people in various environmental and lighting conditions. The training datasets for facial recognition models in most countries are quite limited because this tends to be privacy-related, with natural consequences of not performing as they should under real conditions, such as crowded places or low light.

The challenges will be overcome by establishing partnerships with organizations that hold the relevant datasets and may also make significant investments in the creation of synthetic datasets to help train models of AI. Such proper anonymization techniques must be used so that privacy laws are respected while preserving the utility of the data.

2.2.2 Model Interoperability

Model interpretability is going to be another very challenging factor in integrating AI into digital forensics. AI models, such as deep learning and neural networks, have often been described as "black boxes" because many decisions made by the model may not be easily understood. Interpreting some of the more complex AI models is an important obstacle in deploying AI in digital forensics, especially in legal contexts.

2.2.2.1 Black-Box Nature of AI Models

The black-box nature of AI models denies, by definition, investigators all access to a complete understanding of how a model arrives at its conclusions. Many AI models, especially deep learning-based ones, base their computation on highly complex algorithms for manipulating data. These can predict or classify based on patterns in data that may not be analytically separable by a human expert [5, 6]. In forensic investigation, this lack of transparency is a problem, as investigators need to explain how and why any particular conclusion was reached – and especially in contexts where that conclusion triggers legal consequences.

For instance, in the event that AI is supposed to presume the probability of a given individual reoffending, then investigators – including the court – should be able to know what rationale there is as to why the AI arrived at that particular result [7]. In most cases, AI models – especially predictive models – are complicated by their opaqueness, such as in the case of the COMPAS case, which was criticized based on the AI model's biases on judicial risk assessment. If the model cannot explain why it singled out a particular person as high risk for reoffending, then all the conclusions drawn by the model may have to be questioned for their validity and fairness in arriving at the judgments. In the same vein, AI-driven systems applied in digital forensics to detect fraud or cybercrime have to be interpretable so that results gained can be defended in a court of law by the investigators.

One significant concern with the black-box nature of AI models is the risk of introducing biases or errors that are difficult to detect. If an AI system makes a decision based on biased data or an incorrect assumption, the lack of transparency may make it hard for investigators to identify and correct the issue. This is especially problematic in legal contexts, where fairness and accountability are paramount.

2.2.2.2 Legal and Ethical Implications of AI Decisions

[8] AI models are such that they are black-boxed. The features that factor into the decision-making process of AI systems concerning criminal cases affect the outcomes in criminal cases and may result in unequal punishment pronounced during judicial procedures. This essentially affects ethics as it questions accountability, fairness, and justice. As the reasoning behind those decisions cannot be known, people still question accountability, fairness, and justice.

[5–9] Using facial recognition in digital forensics as an example, if an AI model incorrectly identifies a suspect, it could lead to an arrest or surveillance wrongly. If the investigator fails to explain the pathway by which the AI system reached its conclusion, such a judgment would make it difficult to decide whether this is the right decision or if it is an "error" introduced into the system. In such cases, unfair penalties may have to be borne by people as the system is not transparent during its decision-making.

[10] Ethically, the biggest concern is that AI systems might perpetuate the biases inherent in the training data as this is a serious issue with digital forensics when crime has been identified or even in predicting future behavior. If such training data happens to have a bias perhaps overrepresented by certain groups or scenarios, the AI model could focus on an unfair proportion of individuals based on race, gender, or socioeconomic background.

To alleviate some of these concerns, AI models in digital forensics must be designed with explainability in mind to allow investigators and legal professionals to know or understand how decisions were reached, especially when those decisions directly affect the rights or freedoms of individuals. Models also must be audited regularly in terms of fairness and accuracy, and biases must be notified for correction.

- Legal and Ethical Implications: Uninterpretable AI decisions may contravene the requirements of due process in courts.
- Impact on Investigator Trust: Tools that the forensic professional cannot understand or control may make them resistant to adopt them.

2.2.3 Scalability Issues

With the increasing intricacy of digital forensic investigations, growing data volumes and types also become coupled with the analysis requirements. One of the predominant strengths of AI is the scalability of processing large amounts of information, amounting to what would be overwhelming for traditional forensic methods. In turn, however, scalability presents many challenges, as people must alter some conceptions to successfully integrate AI in digital forensics. The following are some of the challenges: managing huge amounts of data, real-time processing for probe investigations, and so on. The following section elaborates on these issues.

- Managing Big Data: Researchers may work with terabytes' worth of data generated from various devices and networks. AI tools have to scale in a practical way to handle such workloads.
- Real-Time Processing Needs: The demand for real-time analysis, especially during incident response scenarios, requires powerful AI systems that are designed to handle streaming data.

2.2.3.1 Handling Large Volumes of Data

Volume: Data can be enormous, and cyberattacks, fraud investigations, or large-scale data breaches can easily reach the level of terabytes or even petabytes of data. Data here comprise emails, social media posts, network logs, videos, images, or system files. On the one hand, AI is good at handling large volumes of data. On the other hand, its ability to handle enormous amounts of data presents significant challenges related to storage, computing power, and algorithms.

One of the major limitations of such massive amounts of data is that it overwhelms the human-based forensic investigator who depends on manual processes. By automating the process, AI can classify, categorize, and identify any relevant evidence that lies hidden in this sea of data. For example, algorithms developed by machine learning can sift through log files to detect anomalies or suspect patterns in network traffic that would otherwise go unnoticed. Artificial intelligence tools can scan millions of transactions when detecting fraudulent behavior based on learned patterns as compared to financial crime investigations.

However, handling such an enormous amount of data necessitates large computing resources. The storage requirement of such an amorphous dataset is humongous, and without proper infrastructure, it may turn out to be a bottleneck for storing and processing the data. Cloud-based solutions present one possible answer, scaling up available storage for forensic investigations according to need. Other ways that deal with the load include the use of distributed computing techniques and parallel processing by splitting the data-processing load across multiple systems. The management of these, though, is technical in nature.

For instance, in the case of a more comprehensive investigation of a compromise, an AI system may be required to trawl through network traffic logs from multiple servers often geographically dispersed. This necessitates effective data management systems that can pull data from various sources, process it in parallel, and then correlate findings. It's important that it can scale to meet these needs without sacrificing the quality of its analysis.

2.2.3.2 Real-Time Processing Needs

Another challenge that is real time, facing the integration of AI in digital forensics, involves processing information in real time. In many cases, especially those dealing with cybersecurity and law enforcement, the ability to process and respond in real time is critical. An example may be a live cyberattack, where forensic investigators must monitor network traffic as the attack is happening. In such scenarios, real-time anomaly detection and pattern identification may prevent further damage or identify the perpetrators.

[11] However, real time has major constraints in AI systems. The algorithms for AI, especially deep learning and more complex statistical models, are typically computationally resource-intensive. A primary technical challenge, therefore, is to ensure that the AI system can inform in real time with large volumes of data.

Such AI-powered tools in cybersecurity are needed to process data as it continuously flows into a system. For example, AI tools that monitor the activity of users over a network must continually analyze every action in real time to detect any anomalies in the system, such as unusual login times and unauthorized transfers of data. The challenge here is not only in volume but also in processing speed. If it's taken slowly, it can cause misses in certain detections, and in response time, the attackers would be able to infiltrate or exfiltrate data.

For instance, a more practical application of such a problem is the utilization of AI for the purpose of establishing a real-time ability to detect distributed denial of service attacks. The DDoS attack overloads a server with traffic, leaving it entirely incompetent to work. An AI system needs to analyze incoming traffic to identify patterns of an attack, let alone block malicious requests, and still allow legitimate traffic to pass through. Making this occur without delay is an extremely demanding task, requiring high-performance computing resources and highly optimized algorithms.

Moreover, integration with AI for real-time analysis would necessitate edge computing, where the data is processed closer to its source, meaning nearer to the device itself, rather than transmission to a centralized server. It reduces latency, with the responses coming faster. In digital forensics, edge computing can help gain insight into suspect devices – wireless phones or IoT devices – directly onsite, without waiting for all the data to be uploaded to the cloud for analysis.

In total, for all the benefits that AI can afford to handle massive amounts of data with real-time analysis, there is a complex challenge involved in ensuring that the AI system scales efficiently and gives timely insights. It will require additional advanced computing resources, distributed systems, and highly optimized algorithms.

2.3 TECHNICAL LIMITATIONS

2.3.1 Lack of Standardized Tools and Frameworks

The lack of standardization in tools and frameworks is one of the key challenges when integrating AI into digital forensics. Forensic investigators typically deal with a diverse array of tools, as they focus on particular kinds of evidence, for example, file systems, network logs, or encrypted data. Although AI-based tools possess more advantages, they are still not uniformly deployed in these existing systems, and their development does not follow strict uniform standards.

As such, the non-standardized AI tools and frameworks make it difficult for forensic professionals to apply AI in varying cases and on different platforms. As each AI tool uses different algorithms, interfaces, and workflows, investigators would have to learn and adapt to different systems. Moreover, as digital forensics with AI is still an emerging field, most of the AI tools are in their experimental stages, with some levels of unreliability and inaccuracies.

For instance, the number of proprietary forensic tools available for analysis of digital evidence found in mobile devices is numerous, but very few of these tools standardize AI. An investigator might have to use multiple, disparate tools to analyze different types of data from the same device, which further complicates the forensic process, increases the risk of errors, and also reduces the efficiency of the investigations as a whole.

Moreover, the absence of standardized frameworks for AI in digital forensics proves to be problematic when considering interoperability. The more AI-driven tools are integrated into the forensic ecosystem, the greater the need for these tools to work seamlessly without any issues. If the AI tools cannot communicate with each other or share data in a standardized format, investigators will have to manually transfer information between systems, leading to errors and slowing down the whole investigation.

Standardized frameworks of AI should be developed and used in digital forensics with concerted efforts. This way, all the tools developed based on AI can be easily assimilated into the forensic workflow because they would use a common language and structure. Industry-wide standards for data formats, security protocols, and AI algorithms will also streamline the whole forensic process in determining reliability in the AI-based analysis of evidence [7].

2.3.2 Integration with Existing Digital Forensic Tools

Another technical limitation is in the integration of AI with existing digital forensic tools. Most current forensic tools are effective, yet they were not built with AI technologies in mind; therefore, integrating AI may mean that there is a need for significant modification or, in some cases, entirely new systems.

For instance, traditional forensic tools that deal with file carving or disk imaging are meant to extract files from digital devices. Yet, these tools may lack the capability to process AI models in the recognition of patterns and the detection of anomalies or other data analyses. To effectively implement AI, the investigator might need either to add new capabilities to already existing tools or to develop new tools that can operate on both traditional and AI-based evidence.

The process of AI integration also pertains to compatibility issues with legacy systems. Most organizations use established forensic tools that have been in use for quite a while. These systems may not be equipped with the hardware or software capabilities necessary to run AI models, especially those that require powerful GPUs or specialized processors. As AI technologies continue to evolve, the gap between traditional and AI-driven systems will likely widen, making it more challenging to integrate AI into established workflows.

Furthermore, data silos may also be another effect of AI integration; that is, the output of AI systems may not be easily shared or accessed by other forensic tools. For instance, AI may flag certain files as suspicious; however, if such files are not analyzable further by traditional forensic tools such as password cracking and data recovery tools, it might slow down the investigation.

The challenge would have to be addressed by ensuring that digital forensics professionals are competent in training and upskilling, familiar with both traditional tools and AI-based systems. Forensic software developers must focus on developing AI-compatible tools, allowing them to seamlessly operate side by side with legacy systems. There would be a critical collaborative role that tool developers and forensic experts would have to play along with the AI researchers in this space.

2.3.3 Dependence on Computational Resources

AI-powered digital forensic tools are often computationally intensive, requiring substantial processing power and storage. This dependence on high-end computational resources can be a significant barrier, particularly for smaller forensic teams or organizations with limited budgets.

[12] A modern forensic AI model, such as the many deep learning models, requires considerable computational power to train and deploy effectively. Training a deep learning model requires repeatedly processing large datasets, which may take days or even weeks on traditional hardware. Furthermore, when trained, running the AI algorithms on new data for analysis can further require additional processing power, especially in the case of significant amounts of evidence.

All these have to be computed often, and forensic investigators must depend on cloud computing resources or even distributed systems for such needs. Though cloud computing allows for scale and flexibility, it also poses problems regarding the security, privacy, and accessibility of the data. For example, when uploading confidential forensic data to the cloud for processing, it becomes vulnerable to instances of breaches or unauthorized access that threaten data safety and legal standards compliance.

Another issue is that high computational requirements for AI-driven tooling can slow down the overall forensic process. Investigators may sometimes experience a delay in receiving results from AI-based systems, especially when dealing with large-scale investigations based on large volumes of data. The cost of implementing AI in digital forensics may also be prohibitive, as it requires specialized hardware such as GPUs and TPUs.

To overcome this limitation, forensic teams may have to explore alternative solutions, such as hybrid cloud models, edge computing, or optimization of hardware.

AI models can also be designed to be more efficient and less computationally intensive. Forensic organizations should then calculate the cost of AI tools; the infrastructure needed for their execution must be considered in planning digital investigations.

2.4 ETHICAL AND LEGAL CHALLENGES

Therefore, the integration of AI with digital forensics raises possibilities for groundbreaking work simultaneously amid fairly complex ethical and legal issues. For instance, with AI, more complex data can be analyzed instantly and accurately, but again, the question arises regarding the application and involvement in legal contexts. The three core areas of challenge in the mainstream face of forensic professionals are privacy issues, potential biases, and the possibility of AI-generated evidence being declared inadmissible in a court of law. All these concerns raise much more complex questions regarding the technical implementation of the tools, rather than throwing light on the broader implications of AI in the justice, fairness, or accountability case in the legal system.

2.4.1 Privacy Concerns

[5] The most salient ethical issue regarding AI applications in digital forensics is the invasion of privacy. Every AI tool has the potential to siphon through a huge amount of personal data, and the chances of misuse or accidental exposure of the information thus collected create a significant concern. The more entrenched AI is within the investigation process, the greater the danger of infringement on privacy rights, particularly when dealing with inherently sensitive data like communications, financial records, or health records.

For instance, during a cybercrime investigation, law enforcement may rely on AI for analysis of personal communications as well as metadata and browsing habits. This can uncover significant confidential information, making individuals subject to scrutiny from unscrupulous elements. Using AI in monitoring social media handles or online communications has gained notoriety as a privacy violation. For example, facial recognition technology was applied to track the movements of individuals through public spaces and stirred up debate over the need to balance the greater good, public safety, with individual rights to privacy. AI-based surveillance systems installed in an airport, a city, and other public places can take pictures of people without the knowledge of the individuals.

One true example of privacy relates to an application area of AI in surveillance, whereby an artificial intelligence-based surveillance system is applied by police departments throughout the US. For example, the city of San Francisco banned the use of facial recognition technology by government agencies in 2019, becoming one of the first in the United States to do so. The concern is largely about privacy rights and how AI systems may be employed in ways that disproportionately target communities from minority backgrounds.

AI tools used in data analysis in the digital forensic context should be designed with privacy infringement to a minimum. One way to mitigate these issues is through privacy-preserving AI technologies, such as differential privacy, that enable organizations to use and analyze data without revealing specific information concerning individuals. The federated learning technique, where AI models are trained locally on devices rather than in centralized servers, also minimizes the need for data sharing, hence reducing the risk of violations of privacy.

- Issues with cross-border data: Studies that go across jurisdictions raise problems of privacy and compliance with laws such as GDPR.

2.4.2 Bias and Discrimination in Artificial Intelligence Models

Another particularly serious ethical issue with AI in digital forensics is bias and discrimination. AI models are typically trained on historical data that reflect historical societal inequalities, which are unfortunately internalized in bias. Thus, if such biases are not identified and addressed, AI systems may aggravate or perpetuate existing prejudices and consequently culminate in unfair and discriminatory forensic investigative outcomes.

AI algorithms, particularly in domains such as facial recognition, predictive policing, and credit scoring, have been illustrated to display racial and gender biases. [10] The most common of these examples comes from Joy Buolamwini, a researcher at the MIT Media Lab, who discovered that commercial facial recognition systems tend to misclassify more faces of women and people with darker skin tones. This was due to the disproportionate representation of white male faces in the training datasets used to build these AI models.

Digital forensics also makes something amiss where biased AI systems can lead to wrong accusations or false identification based on race, gender, and other social characteristics. For example, [9] a ProPublica study released findings that the US criminal justice system-compelled risk assessment tool called COMPAS was biased in flagging black defendants as at higher risk of reoffending compared with whites. This example illustrates how AI models, when trained on biased historical data, can lead to unfair and discriminatory outcomes.

These biases can also, in a latent manner, influence evidence handling during forensic review. Here, the AI-based tools that are understood to be capable of detecting aberrant patterns in financial transactions will mark a greater percentage of those transactions linked to specific ethnic or socioeconomic traits. In such respects, there will be oversurveillance of certain populations and aggravation of the disparities already present in the criminal justice system.

In providing a step to counter AI bias, therefore, digital forensics professionals need to ensure that the AI model version being used is fair and transparent. A practical way is through the active integration of diverse training datasets that reflect all demographics, including race, gender, and age. Moreover, AI models need to be audited continually for bias, with transparent procedures, where the performance of algorithms is trackable over time. Explainable AI, or how to explain decisions made by AI models, is also another important usage scenario that may help reduce bias in the predictions and contribute to a better understanding of how an AI model arrives at a particular conclusion.

Such a fear has led to the development of guidelines concerning fairness and transparency in the development of AI systems by companies such as IBM. Actions such as these lead to fewer risks associated with biased AI models while promoting the responsible application of AI technologies in digital forensic analysis, among others.

2.4.3 Compliance with Legal Standards

Perhaps the most significant hurdle that AI brings when implemented into digital forensic analysis is its legal admissibility. With its supportability in revealing insights and patterns within digital data for investigators, the integration of AI-generated evidence varies significantly between jurisdictions as authorities control their admissibility based on applicable legal standards. It must also be able to produce a piece of evidence that is sound within legal principles and adheres to the concepts of standard forensic procedures.

Generally speaking, most jurisdictions' principles that create legal frameworks around the admissibility of evidence are based on relevance, reliability, and chain of custody. Consequently, regarding AI-generated evidence, reliability and the integrity of tools should be established when collecting and analyzing digital data. A few examples include the United States, under the Federal Rules of Evidence, wherein it requires digital evidence to prove accuracy and to be obtained through lawful means. This means that a tremendous amount of evidence processed by AI requires clear documentation about exactly how AI models were trained, how they work, and how their findings were verified.

[13] For instance, in 2019, in Edwin M. Walker's US court trial, who was indicted for managing a manipulated Ponzi scheme, financial transactions and digital communications were analyzed by AI tools. Attorneys who represented the defense challenged the reliability of the AI tools used to analyze the evidence placed before them. They further held that forensic investigators should, therefore, be able to provide explanations and defend conclusions reached by the AI algorithms in court.

One such aspect of the use of AI in digital forensics is the black-box nature of many models, which may not even be fully understood by their creators. Predictive policing tools, for example, which come with AI algorithms that predict where crimes are most likely to occur, are a very good case. The opacity of AI's decision-making becomes an issue in such cases. The opaque nature of the model makes it difficult for forensic investigators to justify conclusions drawn by AI before the courts.

Therefore, the AI tools to be applied to computer forensic operations should be designed with consideration of the ancillary legal context. Indeed, the algorithms used for the collection, analysis, and presentation of evidence should also be explainable in terms understandable to a judge or even a jury. Regular audits of AI systems in combination with verification methods can ensure that the AI works as intended and that its inferences are correct and legally reasonable. Third-party audits by independent experts further add credibility to AI systems applied in digital forensic applications in some cases.

2.4.3.1 Variations across Jurisdictions

Complexity surrounding the jurisdictional issue is one of the major challenges to deploying AI in digital forensics. Digital evidence can readily cross over into different countries with their laws on data collection and processing while ensuring all evidence collection, which brings different levels of policing authority. The widespread recent growth of cloud computing, coupled with cross-border data flow, has made the process of passing through the legal minefield in international investigations extremely difficult.

In a few instances, these AI systems used in one jurisdiction are actually not legal enough for another. For instance, using specific AI tools for accessing and analyzing encrypted data may be legally allowed under some forms of national security laws in one country but may be considered a forbidden act or violation of privacy and data protection laws in another country. Moreover, other jurisdictions have their own laws regarding where the data may be stored in the cloud, leading to confusion about what is permissible. This calls for international agreements and treaties over standards on how AI can be used in global digital forensic investigations.

Transnational AI-based investigations need effective conflict of standard law management. International work, such as the Budapest Convention on Cybercrime, has advanced because of the steps towards standardized digital forensic processes. However, most of the AI-driven tools are well beyond the legislative standards, which have created holes in the regulatory framework. Both investigators and organizations will have to ensure compliance at both local and international levels, with probable negotiations entailing the permissions or agreements over access or data sharing when crossing borders.

[14, 15] A specific case illustrating the challenges posed by jurisdictional differences can be seen in the Microsoft Ireland case (2013–2018), in which the US government sought access to emails stored on a server located in Ireland as part of a criminal investigation. The case raised complex questions about the applicability of US laws to data stored overseas and highlighted the potential conflicts between privacy laws in different countries. The more meaningful legal challenges would be in the scenario of applying AI to cross-border analysis and processing of digital evidence.

2.4.3.2 Admissibility of AI Evidence in Court

Another excellent challenge is the admissibility of AI-generated evidence in court. Courts in virtually every jurisdiction have been very cautious about admitting AI-driven conclusions without such insight into how the AI arrived at its findings. An exemplary case of such an issue can be seen in the Daubert v. Merrell Dow Pharmaceuticals case, decided in 1993 by the US Supreme Court. Scientific evidence has to be both relevant and reliable to be admissible in court, the Supreme Court has decreed. These court decisions serve as the current underpinning for the use of expert testimony in US courts. The principles will also be applicable to digital evidence like AI-generated conclusions.

Forensic evidence preferably should be reliable, consistent, and transparent for courts. However, AI models, especially deep learning and other non-transparent complex algorithms, operate in ways not easily understandable or even known to their creators. This "black-box" nature can lead to questions about the validity and reliability of AI-generated evidence.

Thus, for AI evidence to be admissible in court, it has to pass both scientific rigor and the legal requirements of chain of custody, transparency, and reproducibility standards. There are likely to be higher expectations of the forensic investigators when explaining what the AI system produces its results based on, how the data were processed, and what possible errors or uncertainties may exist in the system. Explainability is critical to enable full understanding by juries and judges of any AI-based analysis results and to assess the credibility of those results.

Secondly, before use in forensic investigation, AI tools must undergo rigorous validation and testing. Validation spans assessing that AI systems give repeatable results across different conditions through proper checks. The tool must also demonstrate reliability across different kinds of digital evidence, such as emails, documents, images, and network logs.

For example, when AI is used in detecting and preventing fraudulent financial transactions, the forensic expert may be required to describe what kind of training data was used, what algorithm was used, and what validation process assures that the AI output was valid. Courts will summon forensic experts to prove that the AI tool applied to the evidence analysis meets specific admissibility standards within the court, these criteria being reliability, transparency, and accuracy.

As this technology develops, courts will be asked to rely on experts in AI and digital forensics to explain sometimes very complex results and corroborate the reliability of AI-driven tools. Thus, there's a stark need for a change in the skill sets of those legal professionals who would like to see an increasing demand for cross-disciplinary collaboration between legal experts, forensic investigators, and AI specialists.

In the future, we shall see the emergence of expert tools developed by AI targeting court admissibility with explainable AI techniques and standardized validation procedures to ensure legal compliance.

2.5 SECURITY RISKS IN ARTIFICIAL INTELLIGENCE-BASED FORENSICS

2.5.1 Vulnerabilities to Adversarial Attacks

Digital forensics-based AI systems largely face adversarial attacks. Adversarial attacks leverage subtle input data manipulation and attempt to find a way to fool AI models into predicting or classifying incorrectly. In digital forensics, an adversarial example can change digital evidence, such as an image, audio, or malware, in such a way that an AI-based forensic tool may fail to identify some critical evidence. [1] Adversarial perturbations could be so subtle that even a fraction of the elements used to identify an image, for instance, facial features when identifying criminals, might be altered to either misidentify the image or be insensitive to significant elements within the image. The case is similar in AI models used for malware, where slight modifications to the code of the file could be imperceptible, such that the AI will classify the benign file as a malicious file.

The major problem with adversarial attacks is the "black-box" nature of most AI models. These models, especially deep neural networks, have learned complex patterns from data but do not give users any clear insight into how they arrived at their decisions. In this way, they prove opaque about the occurrence of an attack and how to respond.

There are several techniques proposed to counter adversarial risks. Among the best techniques, adversarial training, which is based on exposing the AI system to adversarial examples during training such that the model can learn how to identify and resist such attacks, is very prominent. Robust optimization-based approaches, like defensive distillation and feature squeezing, are used to suppress the vulnerability of the model to adversarial manipulations. [16] In forensic investigations, it would be necessary to use XAI that can explain the reasoning of an AI model's decision-making. This way, knowing how human expertise is combined with insights based on AI in a manner that minimizes the possibility of adversarial attacks means that forensic investigations are sound and valid.

2.5.2 Risk of Artificial Intelligence Misuse by Cybercriminals

AI technologies, while offering significant benefits to digital forensics, also create opportunities for cybercriminals to exploit and misuse these tools. Cybercriminals can use AI to automate and enhance their malicious activities, making them more difficult to detect and counter.

For instance, phishing attacks can be conducted using AI-driven tools. AI systems can generate messages through personalized phishing emails or social media, increasing the chances that the scammer will be successful. Additionally, AI can create deepfake videos or audio, which cybercriminals can then use to manipulate evidence in investigations related to criminal cases. Therefore, in the context of digital forensics, AI might be used to create or obscure digital evidence, making it draw wrongful conclusions or undermine investigations.

In addition, cybercriminals can use adversarial AI models to evade AI-based detection systems. They may modify malicious files or activities in ways that can bypass AI tools established to detect them. For instance, AI-based malware detection systems may be tricked into considering a piece of malware harmless by just adjusting its signature slightly, thus making the system useless [17]. This can be combated through the establishment of AI security measures: robust anomaly detection, secure model training, and the implementation of integrity checks. In addition, joint efforts by law enforcers, cybersecurity professionals, and other researchers are necessary to share knowledge and develop anti-AI-powered crime workarounds [18].

2.6 HUMAN FACTORS

2.6.1 Skill Gap among Digital Forensics Professionals

A considerable skill gap for forensic professionals has appeared as the use of AI becomes more prevalent in the process of digital investigation. The traditional forensic expert and practitioner are not technically equipped to correctly understand, implement, and interpret the deployment of AI-driven tools and methods. This is primarily because the rate at which AI is changing exceeds that of most knowledge-changing events, and a number of the complexities in most machine learning models require specific knowledge of a set of subjects such as data science, programming, and statistics.

The skills gap is also likely to affect the efficiency and reliability of forensic investigations. Forensic experts will understand neither the limitations of an AI tool nor the possible misinterpretation of results, failing to reap the full benefits of AI capabilities. Additionally, the unfamiliarity of AI technologies may lead to resistance to using them.

To fill this gap, comprehensive training programs need to be developed for the technical as well as forensic aspects of AI. Academic and professional organizations should design courses, certifications, and workshops for training digital forensics practitioners on AI-related concepts, algorithms, and practical applications. Interdisciplinary research between AI researchers and forensic practitioners will greatly help configure AI tools according to the specific needs of digital forensics, making them user-friendly and accessible.

2.6.2 Resistance to Adopting Artificial Intelligence Tools

Another significant challenge is the resistance to adopting AI tools in digital forensics. The resistance can be due to a variety of reasons, including distrust in the reliability of the AI models, job displacement anxiety, and ethical implications surrounding the use of AI during forensic investigations. Forensic experts are accustomed to traditional methods and may be apprehensive about trusting automated systems that lack human oversight and judgment [19]. One of the major concerns is that many AI models tend to operate in a black-box nature. Forensic experts are unlikely to be pleased with the idea of relying on AI tools whose decision-making processes are opaque. Lack of interpretability raises serious concerns regarding the fairness and accountability of AI-driven investigations, especially if the results tend to find their way into legal processes as evidence.

The increase in awareness of the advantages of AI in digital forensics, such as handling large volumes of data, automating time-consuming work, and adding precision to investigations, can overcome this resistance. Additionally, the advocacy for developing an XAI model that not only delivers clear explanations but also results in decisions taken by AI can establish trust among forensic experts. It is also important to emphasize that AI tools augment human expertise and do not replace it; even human oversight is required to ensure the ethical and legal validity of forensic finding. A proper balance between automation and human oversight must be struck in digital forensics. AI is excellent in processing big datasets and finding patterns, whereas some judgment regarding its applicability and admissibility in court must necessarily be made by humans. [20]

2.6.3 Balancing Automation and Human Oversight

Striking the right balance between automation and human oversight is critical in digital forensics. While AI excels at processing large datasets and identifying patterns, human judgment is indispensable for interpreting findings and ensuring their legal admissibility.

Best practices include designing systems with built-in human oversight mechanisms, such as manual review stages. Forensic teams should work closely with AI developers to ensure that tools meet both the goals and the ethical standards of an investigation.

Whereas AI will strengthen the efficiency of digital forensics through automation of mundane tasks, there is a tremendous need to strike a balance between automation and human oversight. While AI systems are capable of doing large data processing and pattern recognition, most of their strengths are lacking in the nuanced understanding that a human brings to the table. Human oversight is the only way to ensure that AI generated insights can be interpreted correctly and that ethical standards are maintained.

For instance, AI-based tools can be used to analyze digital evidence, identify potential threats, and flag anomalies. However, a human forensic expert must review the context of the findings, make informed judgments, and then decide on the relevance of the evidence in a legal or investigative context.

There must be a distinct framework set up for the collaboration of AI with humans. This framework must clearly spell out the roles and responsibilities of both the human expert and the AI system. AI tools will be used in this regard to enhance work efficiency by running repetitive tasks and analyzing large volumes of data. Humans can now be responsible for decision-making and ethical considerations, especially because digital forensics will remain reliable, transparent, and legal.

2.7 CASE STUDIES AND EXAMPLES

2.7.1 Successful Artificial Intelligence Implementations in Digital Forensics

AI has been successfully applied in various forensic domains. For instance, AI-based image recognition systems help identify child exploitation material. In this regard, these systems have greatly reduced the burden on investigators. To facilitate fraud investigations, natural language processing tools analyze chat logs and emails to unveil evidence.

There have been significant successful implementations of AI in digital forensics, demonstrating the transformative potential of these technologies. Such an example is the use of AI in the identification of child exploitation through image recognition. AI-based tools can auto-scan large datasets of images and videos to automatically classify content as illicit, saving much time for forensic analysts.

In terms of cybersecurity, AI systems have been applied to the detection of malware, identification of vulnerabilities, and prediction of potential attacks. Such tools apply machine learning algorithms to recognize patterns in network traffic or file structures, allowing them to learn to recognize unknown threats. All of this makes AI accelerate and improve the quality of cybersecurity investigations.

Moreover, NLP is used to analyze textual evidence. These include chat logs and emails in any kind of fraud or cybercrime investigation. AI-based NLP systems can extract relevant information; they can identify relationships between entities and look for latent patterns to help investigators make crime identification more accurate.

These indicate that AI can complement traditional forensic procedures and improve the efficiency of an investigation. This calls for dealing with several problems introduced by the integration of AI, such as data quality, model interpretability, and human oversight.

2.7.2 Failures and Lessons Learned

Although promising, AI in digital forensics has sometimes failed to achieve what its proponents expected. Bias or incomplete datasets are often blamed for some failures. Example: AI algorithms trained on biased or unrepresentative data may generate inaccurate and discriminatory outputs. A common example was the facial recognition system used for forensic investigations, where an AI-powered facial recognition system was found to have higher rates of error for certain demographic groups, resulting in wrongful identification.

Lastly, the utilization of AI when attempting to predict criminal behavior was a total failure. In some contexts, the AI models looked entirely at historical data that contained numerous biases and systemic inequalities, meaning there was an unfair prediction as the AI calculated the chances of an individual committing crimes.

These failures make it imperative to test, validate, and provide transparency in AI systems. It is therefore important that AI tools are trained on diverse, representative datasets and that their outputs are tested appropriately before deployment in real forensic investigations.

2.8 RECOMMENDATIONS AND BEST PRACTICES

2.8.1 Strategies for Enhancing Data Quality

Ensuring data quality is critical to effective AI integration. Strategies include standardizing data formats, implementing robust data cleaning protocols, and using synthetic data to augment training datasets. Techniques like data augmentation and federated learning can address the problems related to data scarcity and privacy.

Improving data quality is key to ensuring the effectiveness of AI in digital forensics. Some of the strategies for achieving data quality include data cleansing, which refers to removing noisy, incomplete, or irrelevant data before it is used to train AI models. This may make the inputs of the system more accurate and reliable.

Synthetic data has great importance in augmenting real-world datasets. This technique focuses on using sensitive or hard-to-collect data, where techniques like data augmentation increase variations of already existing data, generally improving AI models' generalization.

2.8.2 Frameworks for Ethical Artificial Intelligence Use in Forensics

Therefore, applying AI to digital forensics is very challenging in terms of ethical aspects, such as openness, accountability, and the danger of misuse. The majority of forensic investigations deal with sensitive data, so the application of AI tools should be strictly managed to ensure that they are applied ethically. Several frameworks have been developed and applied to guide ethical AI use in forensics to ensure that the practice of AI adheres to normal standards of integrity and fairness.

One underlying principle of ethically deployed AI in forensic applications is responsibility. AI models are only as good as the data they were trained on, and if such models then make wrong or biased decisions, it is critical that someone be held accountable. In digital forensics, this could mean ensuring that AI decisions can be traced back to human oversight. Explainability of how an AI model came to a conclusion (explainable AI, or XAI) is critical in maintaining accountability. For instance, when AI tools are used for the analysis of digital evidence, forensic experts should be in a position to know how the tool identified specific data as relevant or suspect. The risk is that without explainability, AI decisions may be considered "black-box" judgments; therefore, this could raise potential issues in legal settings where transparency is crucial.

Another principle is fairness, which speaks to the dangers of possible bias within AI algorithms. In the realm of forensic science, such bias could be traced along racial, gender, or socioeconomic lines, especially as algorithms are used in predictive policing or facial identification. Ethical frameworks for AI in forensics highlight the need for diverse and representative datasets to mitigate these biases. For instance, a facial recognition system trained on predominantly white male faces will likely produce erroneous results when used on individuals of other ethnicities or genders. Best practices, such as those outlined by the AI Now Institute, encourage AI models to be subjected to rigorous audits for bias before deployment, and that organizations like forensic labs commit to continuous oversight and adjustments to minimize biases when they appear.

Transparency and privacy are other important factors in ethical AI deployment. While going through personal data during forensic investigations, privacy must not be breached during the analysis of AI. This can include respect for personal data privacy laws like GDPR in Europe or CCPA in California, so that the AI tools work only on legally and ethically viable data. Investigators should also be able to justify which specific data is being used in AI-led investigations and how such data is being protected.

The IEEE Global Initiative on Ethics of Autonomous and Intelligent Systems offers a framework of guidelines applicable within the forensic context. Such a principle applies strongly in forensics as the final ruling very often should be made based on a combination of considerations.

These ethical considerations must be integrated into forensic practices through the creation of adequate institutional policies. This could involve the founding of the equivalent of ethical review boards within forensic agencies to monitor the deployment and application of AI technologies for compliance with ethical standards. In addition, a code of ethics for forensic professionals in the application of AI should be formulated, similar to medicine or law counterparts, outwardly published and developed by professional organizations.

Aligning their responsibilities with the principles of justice, accuracy, and transparency in AI use is essential. In conclusion, ethical application frameworks for AI should strike a balance between innovative creativity and strict adherence to ethical considerations. This includes transparency, accountability, fairness, and privacy in using AI, while AI use is assured not to undermine experts but to support them in their vocation of using these tools.

2.8.3 Training and Upskilling Forensic Experts

The fast development of AI technologies opens up opportunities but equally presents challenges to professionals in digital forensics. With the increasing incorporation of AI into the forensic workflow, it's ever more important for professionals to have the knowledge and skills required to handle the tools effectively, maintaining high standards of accuracy and integrity for forensic investigations. The greatest challenge in this area is that there is a cognitive gap between traditional forensic practices and the technical expertise needed to properly apply AI. Forensic practitioners usually work in law enforcement, law, or computer science, and most of them may not fully understand machine learning, neural networks, or data science. This ignorance may translate to problems in interpreting the results produced by AI, embedding AI products in forensic workflows, and understanding the limitations and biases of AI systems. This means that training programs for forensic investigation need to take on an AI-centric character with a focus on machine learning and techniques of data analysis.

One way to bridge these gaps is through specialized courses and certificates. For instance, universities and professional organizations can provide training courses in AI for digital forensics in areas such as machine learning algorithms, data preprocessing, and model evaluation. Not to forget, the development of certification programs where forensic experts obtain recognition in the form of qualifications in AI applications, much like certifications for network security or cloud computing.

Besides formal education, hands-on training is also important. Forensic experts should have the chance to work with AI tools in controlled environments where they can practice the application of AI models to real forensic scenarios, such as simulations of digital investigations using AI tools for tasks like image recognition, data correlation, or anomaly detection. Such programs would provide forensic experts with access to diverse datasets, AI models, and simulation environments, enabling them to gain hands-on experience and develop expertise in applying AI to real-world investigations. Forensic professionals also need continuous professional development (CPD) in a rapidly changing environment like AI. AI technologies change at a rapid pace, and forensic professionals need to keep up with the latest advances, tools, and techniques. CPD programs should encourage forensic experts to frequently participate in training sessions, webinars, workshops, and conferences on AI in forensics. These activities may be under the coordination of professional organizations, academic institutions, or government agencies and should be oriented towards both the technical and ethical usage of AI in forensics.

Interdisciplinary collaboration in training programs also needs emphasis. The application of AI in digital forensics calls for a convergence of data science, computer programming abilities, legal expertise, and awareness of ethics. Forensic practitioners need to collaborate with AI researchers, data scientists, and legal experts to provide an integrative understanding of the impact of AI tools in forensic investigation. Collaborative training programs by experts from diverse fields may assist forensic professionals in putting AI into the overall context and adopting a holistic approach in their work.

Lastly, ethics training is an integral part of upskilling the forensic expert in AI applications. The psychological impacts of AI in digital forensics are associated with ethical dilemmas. Therefore, professionals have to be trained on dealing with issues such as data privacy, algorithmic bias, and transparency. The training should dwell on the ethical and legal standards governing the use of AI in forensics and its potential consequential lapses of ethics.

2.9 CONCLUSION

Effective training and upskilling of forensic experts in AI are necessities for the successful integration of AI technologies in digital forensics.

BIBLIOGRAPHY

[1] Ian J. Goodfellow, Jonathon Shlens, and Christian Szegedy. Explaining and harnessing adversarial examples. *arXiv preprint arXiv:1412.6572*, 2015.

[2] Jo Feller and Sandeep Misra. Exploring the ethical and legal implications of AI in digital forensics. *International Journal of Information Technology & Decision Making*, 20(6):1055–1072, 2021.

[3] Darren Quick and Kim-Kwang Raymond Choo. Impacts of increasing volume of digital forensic data: A survey and future research challenges. *Digital Investigation*, 11(4): 273–294, 2014.

[4] Matt Baker. AI and big data in cybersecurity: Ransomware investigation delays due to scalability issues. *Reuters Technology*, 2021.

[5] Surya Mattu Julia Angwin, Jeff Larson and Lauren Kirchner. Machine bias: There's software used across the country to predict future criminals. and it's biased against blacks. *ProPublica*, 2016.

[6] Drew Harwell. AI and the criminal justice system: Transparency and bias in recidivism predictions. *The Washington Post*, 2019.

[7] Aman Jain. Enhancing forensic analysis of digital evidence using machine learning: Techniques, applications, and challenges. *IJRMPS*, 12, 2024.

[8] Matthew Salzano et al. AI and advocacy: Maximizing potential, minimizing risk. *School of Communication and Journalism Faculty Publications*, 2024. https://commons.library.stonybrook.edu/soj-articles/4

[9] Nigel Stobbs Mirko Bagaric, and Dan Hunter. Erasing the bias against using artificial intelligence to predict future criminality: Algorithms are criminality: Algorithms are color blind and ne e color blind and never tire. *University of Cincinnati College of Law Scholarship and Publications*, 88:1037–1081, 2019.

[10] Buolamwini Joy and Gebru Timnit. Gender shades: Intersectional accuracy disparities in commercial gender classification. *Proceedings of Machine Learning Research*, 81:77–91, 2018.

[11] T. Zhu et al. General, efficient, and real-time data compaction strategy for APT forensic analysis. *IEEE Transactions on Information Forensics and Security*, 16:3312–3325, 2021.

[12] Razaq Jinad, Khushi Gupta, Ecem Simsek, and Bing Zhou. Bias and fairness in software and automation tools in digital forensics. *Journal of Surveillance, Security and Safety*, 5:19–35, 2024.

[13] Dan Sullivan. Real-time data processing struggles in financial fraud investigations. *Financial Times*, 2020.
[14] Sharon Bradford Franklin. The Microsoft-Ireland case: A supreme court preface to the congressional debate. *Lawfare Media*, 2018.
[15] Krunoslav Antoliš. The challenges of collecting digital evidence across borders. *Police and Security*, 32, 2023.
[16] Nicolas Papernot, Patrick McDaniel, Ian J. Goodfellow, Stephen Horn, Z. Berkay Celik, and Ananthram Swami. The limitations of deep learning in adversarial settings. *arXiv preprint arXiv:1511.07528*, 2016.
[17] Nandini Bose and Qingqing Liao. Artificial intelligence for cybersecurity: A review. *arXiv preprint arXiv:1808.06303*, 2018.
[18] Jorge Paz, J. Michael Perez, Sabrina Fink, and Violeta Sanz. Malicious use of artificial intelligence in cybersecurity. *Proceedings of the 2019 International Conference on Cybersecurity and Privacy*, 22–33, 2019.
[19] Marco D. Pereira and Idris Bouchrika. Barriers to the adoption of ai in digital forensics: A qualitative study. *Digital Forensics and Cybersecurity Journal*, 12:45–60, 2019.
[20] Simson L. Garfinkel. Digital forensics research conference: Keynote address on the challenges of digital forensics. In *Proceedings of the Digital Forensics Research Conference*, pages 1–8, Orlando, FL, 2010. DFRWS.

3 Towards Open Standards for Systemic Complexity in Digital Forensics

Paola Di Maio

3.1 UBIQUITY IN DIGITAL FORENSICS

Digital forensics (DF), in the context of forensic science, is concerned with identifying, gathering, storing, analyzing, and communicating digital evidence from electronic devices and computer systems [1]. Its core capability is based on precise information acquisition techniques and logical processing functions, in particular, deductive reasoning.

Sophisticated technologies are now used to support fact-checking and inquiry, not only pertaining to criminal investigations but also in scientific and technical inquiry. Given that in our times all processes are mediated by digital technologies, DF techniques can be used to reconstruct and understand facts and events generated in digital scenarios, as well as in all cases where digital records of non-digital (physical) evidence are used to support legal, scientific, or technical analyses of all kinds.

As such, DF is increasingly relevant to all fields of inquiry and ubiquitous. On the one hand, it consists of processes, techniques, and algorithms applicable to digital technologies supporting the analysis of both digital and non-digital (digital records of physical) evidence, and on the other hand, it consists of machine learning techniques that automate these logical analysis and inferencing processes (Figure 3.1).

As technology evolves, and DF becomes exponentially complex, its principles and techniques become applicable to diverse areas of practice, such as in the responsible management of digital records, digital policy compliance, and reproducible data science. The convergence of artificial intelligence (AI) and DF provides powerful data analysis capabilities, especially the facility to automate the screening of large datasets. Nonetheless, the use of AI techniques in forensics comes with its own challenges and risks that can only be addressed if the underlying logical processes are standardized, transparent, explainable, and reproducible. It is impossible to evaluate the correctness of DF if the data and processes are not standardized or transparent.

This chapter contributes central arguments to the development of open standards in DF to analyze, correlate, and query data stores using explainable AI with the intent to ascertain truth, not only for the purpose of law enforcement but also for organizational governance and scientific inquiry in general, pointing to the need for greater transparency and accountability supported by the human readability of digital and machine learning artifacts.

DOI: 10.1201/9781003512820-3

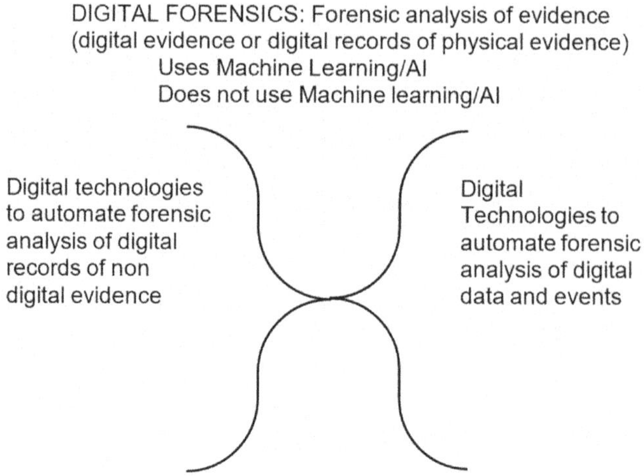

FIGURE 3.1 Ubiquity of Digital Forensics.

Despite the usefulness of novel AI tools, it is ultimately human experts and adjudicators who validate and interpret the outcomes of automated analysis, a practice that is sometimes referred to as humans in the AI loop [2]. The ability of humans to assess the validity of automated processes requires human-readable digital artifacts. To mitigate bias and demonstrate accountability, and as such, the validity of the outcomes of automation, both the data and the processes need to come out of the black box. The approach proposed in this chapter advocates for open human-readable standards in DF.

3.2 SYSTEMIC COMPLEXITY

Thanks to the proliferation of technologies for data storage and analysis, and to widening fields of application, digital forensics (DF) is expanding and now overlaps with other disciplines, leading to a state of **systemic complexity**, defined in terms of its multidimensional evolution and rapid reconfiguration of its functional, systemic, and behavioral boundaries. Technologies based on complex knowledge domains can be powerful yet fragile at the same time. This research identifies this complexity and addresses it with a system-level knowledge representation approach.

It can be said that whatever DF literature typically identifies as forensic evidence can actually be reduced to what computer science refers to as **data**, resulting in an inevitable overlap of DF with CS and Data Science (DS) (Figure 3.2).

Forensic tools and techniques have become necessary to sustain robust inquiry not only in law enforcement but also in every kind of scientific investigation, especially in data science (DS). Due to the exponential proliferation of datasets from research, their size, and heterogeneity as found in typical analytical scenarios, data curation practices require accurate and systematic techniques for the collection, maintenance, and analysis of data, which are instrumental in ensuring the integrity, accessibility, and validity of any finding. For example, data management planning (DMP)[1] methods

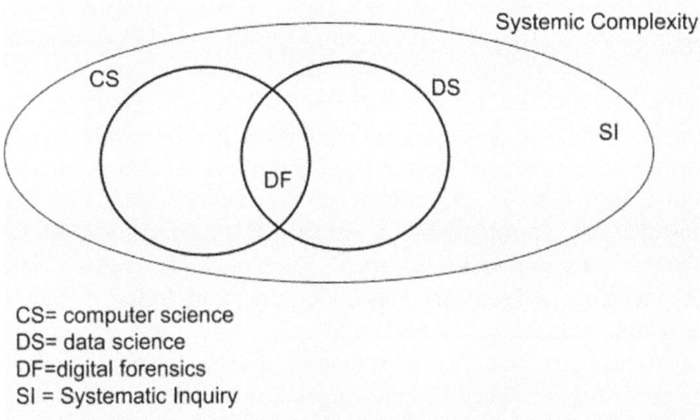

FIGURE 3.2 Systemic Complexity in Digital Forensics.

prescribe how to systematically collect, store, secure, archive, and share data, with their adoption having become a standard requirement in project management and scientific research.

The evaluation of data quality and integrity is central to assessing the accuracy of DF, which also depends on the adoption of well-defined and commonly used data formats for interoperability [3]. Data extraction, pattern recognition, and anomaly detection are DF techniques commonly used in DF and in data science [4].

3.3 DIGITAL FORENSICS AND ARTIFICIAL INTELLIGENCE

Through DF, electronic data can be analyzed in detail to determine the chronological and factual accuracy of information lifecycles, thus enabling their interpretation in support of hypotheses, a logical process of evaluation and analysis that can be enhanced by the use of artificial intelligence (AI).

Early efforts to record and examine individual incidents of user behaviors were ad-hoc and unsystematic; however, together with technological advancement, DF progressively developed towards standardized practices, leading to the development of foundational concepts and to the emergence of a digital forensics community working towards legal frameworks [5].

The continuous evolution of digital forensics aims to achieve stable standardized methodologies and best practices, with emerging specializations from mobile forensics [6] to network forensics [7].

Nowadays, DF advanced technologies are being used for data recovery, analysis, and extraction, and to expose registry information. AI-supported DF leverages machine learning to automate and optimize specific forensic techniques, such as data carving, to reconstruct deleted or fragmented data from storage devices, steganography, stochastic analysis, and file deletion.

Once the data has been identified, cleaned up, and verified, DF tools can be designed to support its use to perform intelligent reasoning and reach conclusions.

Likewise, any misinterpretation and erroneous reconstruction of fragmented or missing datasets can lead to spurious fabrication, justified and supported by the latest technology.

By its programmatic nature, computational environments offer native support to certain analyses, thanks to their systematic processes and protocols, and their inherent logging capabilities. Typical built-in logging utilities that can be leveraged by DF include **system logs**, used to track system start-up, shutdown, hardware events, and software installations; **application logs** that record application activity, errors, and user interactions; **security logs** that capture login attempts, access control events, and security incidents; and **network logs** tracking network traffic, including connection attempts, data transfer, and firewall activity.

Other analytical artifacts that can support forensic analysis include System Resource Usage Monitor (SRUM), Prefetch, and, in Windows OS, ShimCache and *AmCache*, which are *Windows artifacts* that *contain* information about recently executed *applications*.

AI and ML techniques can be particularly advantageous in parsing, analyzing, and correlating data from these diverse sources.

DF architectures can be designed based on logical and functional layers to enable the systematic collection and analysis of data chain of custody to validate assertions of statements of facts in diverse areas of inquiry such as incident response, electronic discovery, corporate investigations, data recovery, and more generally to support evidence-based rational inquiry. AI technologies can be especially useful to facilitate big data analysis [8, 9].

The integration of artificial intelligence (AI) into digital forensics delivers enhanced efficiency in terms of depth and breadth of analysis, as well as speed. AI-powered tools can automate and accelerate exponentially repetitive tasks like data acquisition, processing, and analysis [10], while saving time and resources. By the same token, the rapid automation of analytical processes can enhance the fallacies, flaws, and errors in analytical models and outcomes, exploding the potential rate of error. To identify and correct AI-generated errors, it is necessary to employ highly skilled human experts through manual or semi-manual verification, which can be very costly or even unfeasible at a large scale.

Machine learning algorithms, in principle, can improve the accuracy of results by sifting through vast amounts of data to identify patterns and anomalies, analyze complex relationships between disparate data points that might be missed by human analysts, leading to more accurate findings. Typical applications include the automation of malware and anomaly detection, the analysis of code features and network behavior to identify malicious activities, and predictive analytics, which can forecast potential breaches based on historical data and user behavior patterns [11].

Among the latest ML techniques in use, *memetic algorithms* combine evolutionary algorithms with local solvers referred to as memes [12], and *fronesis* consists of rule-based reasoning and attack detection rules to identify adversarial techniques [13], in addition to deep learning techniques such as "stacked auto-encoders (SAE), deep belief networks (DBN), convolutional neural networks (CNN), and recurrent neural networks (RNN)" [14].

Despite the proliferation of ML techniques and their growing adoption, strong AI is not yet part of the DF process due to innumerable challenges, discussed in the next section, which can contribute to incorrect outcomes, wrong conclusions, suboptimal or wrong decisions, and misjudgment. The challenges to data and process integrity exist in all forensic analyses but can be magnified exponentially when using machine learning.

Research identifies possible approaches to resolve the challenges [15], for example, recommending the use of explicit knowledge representation techniques such as pattern recognition and expert systems to enhance ML capabilities and processes and "ensure that their deployment is ethical, transparent, and respectful of data subjects' rights and privacy" [11].

The chapter contributes directly towards such recommendations.

3.4 OPEN CHALLENGES

The overall challenge and limitation of forensic sciences is the ability to consider appropriate evidence and follow a logical process analysis to reach correct conclusions to support appropriate judgments and decisions. Although DF brings advantages in terms of scale and processing capacity, the use of ML can also weaken the scrutability of processes because of a lack of transparency. A substantial body of knowledge documents how forensic science gone wrong contributes directly to miscarriage of justice (MoJ) [16].

Experts have been calling for "uncompromising mechanisms capable of identifying flaws in the system, instigating appropriate remedial action in particular cases, and ensuring that necessary reforms are implemented to mitigate misjudgement" [17]. The adoption of DF aims to establish verifiable truth of data and processes that ascertain facts and the legitimate adherence to protocols.

Nonetheless, the final validation of the outcomes, to assess their accuracy, integrity, and accountability, still depends entirely on human judgment. DF practitioners typically apply categorical conclusions or strength of support conclusion types that cannot really be verified. Both humans and AI, however, are affected by bias [18]. Forensic science and its practitioners are respected and valued in judicial proceedings due to their capacity to discover irrefutable facts that may otherwise not come to light. But they are not immune from logical fallacies (Figure 3.3) [19].

In particular, confirmation bias is known as "a proclivity to search for or interpret additional information to confirm beliefs and to steer clear of information that may disagree with those prior beliefs" [20]. Blind trust in the correctness of software-generated evidence is known to have resulted in wrongful convictions for decades [21]. There are expectations that AI may help reduce human bias by standardizing the reasoning; however, machine learning-generated bias is a concern [22]. Although scientific methods can be used to assert and evaluate different degrees of confidence and even certainty in the outcome of automated processes, machine learning itself tends to lack an explanation of its meaning or reference to established frameworks.

"Substantial deficiencies in documentation practices for content essential for enabling audit of the DF investigative process and results, a challenge shared with other forensic science disciplines" [23].

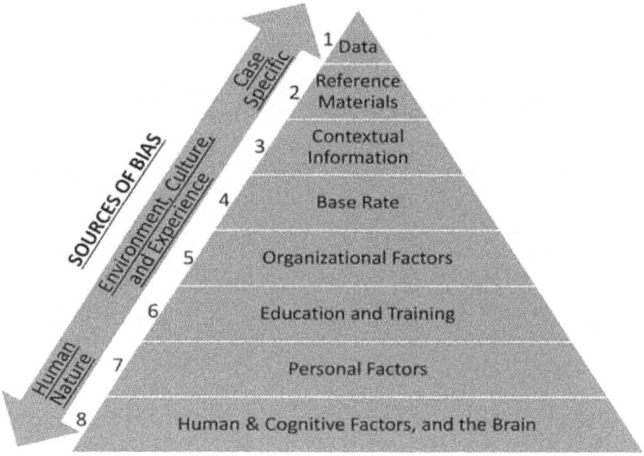

FIGURE 3.3 Six Fallacies and Eight Sources of Bias Dror, I. E. (2020).

The effectiveness of ML models in DF depends on the quality and representativeness of training data. Biases embedded in data can skew results and result in unreliable conclusions. The ultimate logical composition and analysis of the evidence/data identified through the DF process remain a prerogative of the intelligent human expert, capable of establishing the relevance of the evidence, interpreting, and drawing conclusions from the analysis. Understanding and addressing both human and machine bias is non-trivial.

Bias is at the root of all human and machine intellectual and cognitive limitations, and there are countless ways that it can be characterized, starting from philosophical disquisition to mathematical, geometric, and statistical applications [24]. Understanding the reasoning behind AI-generated evidence and decisions remains a challenge, potentially hindering trust and legal acceptance [15].

Integrating AI into forensics raises additional concerns about data security, privacy breaches, and potential misuse of powerful algorithms [25]. Researchers warn that the legal admissibility of AI-generated evidence and the ethical implications of using AI in investigations require careful consideration [26].

There is general acceptance that AI technologies are increasing the efficiency of analyses; however, their reliability is limited due to a lack of transparency [27]. The landscape of challenges for DF is vast and complex. It can be traversed using taxonomic approaches, for example, by identifying categories such as technical, legal, personnel, and operational [25].

USE OF TAXONOMIES IN DF RESEARCH (from Karie Venter)

- Altschaffel et al.: Digital forensic investigations are usually conducted to solve crimes committed by perpetrators and/or intruders using IT systems. They then propose a taxonomy that helps perform a forensic examination and to establish answers to a set of well-defined questions during such examination.

- Hoefer and Karagiannis: Taxonomy of cloud computing services. A tree-structured taxonomy based on their characteristics, to easily classify cloud computing services and make it easier to compare them. In contrast, the proposed taxonomy in this paper offers a simplified platform that sheds more light on the classification of existing digital forensic challenges.
- Strauch et al.: Taxonomy to categorize existing cloud data-hosting solutions.
- Lupiana et al.: Taxonomy for classifying disparate research efforts in ubiquitous computing environments. Their taxonomy classifies ubiquitous computing environments into two major categories: interactive environments and smart environments.
- Sansurooah: Methodologies and procedures that are in place for the gathering and acquisition of digital evidence and subsequently defines which model will be the most appropriate taxonomy for the electronic evidence in the computer forensics analysis phase.
- Sriram: Reviews the research literature since 2000 and categorizes developments in the field into four major categories.
- Kara et al.: New research categories (taxonomy) and areas identified at the Colloquium for Information Systems Security Education (CISSE-2008), as well as a plan for the future development of a formalized research agenda for digital forensics.
- Garfinkel: Adopt standardized, modular approaches for data representation and digital forensic processing.

CHALLENGES IN DIGITAL FORENSICS (Karie and Venter 2015)

1. **Technical Challenges**
   - i. Encryption
   - ii. Vast Volumes of Data
   - iii. Incompatibility among Heterogeneous Forensic Tools
   - iv. Volatility of Digital Evidence
   - v. Bandwidth Restrictions
   - vi. Limited Lifespan of Digital Media
   - vii. Sophistication of Digital Crimes
   - viii. Emerging Technologies and Devices
   - ix. Limited Window of Opportunity for Collection of Potential Digital Evidence
   - x. Anti-Forensics
   - xi. Acquisition of Information from Small-Scale Technological Devices
   - xii. Emerging Cloud Computing or Cloud Forensic Challenges
2. **Legal Systems and/or Law Enforcement Challenges**
   - i. Jurisdiction
   - ii. Prosecuting Digital Crimes (Legal Process)
   - iii. Admissibility of Digital Forensic Tools and Techniques
   - iv. Insufficient Support for Legal Criminal or Civil Prosecution
   - v. Ethical Issues
   - vi. Privacy

3. **Personnel-Related Challenges**
   i. Lack of Qualified Digital Forensic Personnel (Training, Education, and Certification)
   ii. Semantic Disparities in Digital Forensics
   iii. Lack of Unified Formal Representation of Digital Forensic Domain Knowledge
   iv. Lack of Forensic Knowledge Reuse among Personnel
   v. Forensic Investigator Licensing Requirements
4. **Operational Challenges**
   i. Incident Detection, Response, and Prevention
   ii. Lack of Standardized Processes and Procedures
   iii. Significant Manual Intervention and Analysis
   iv. Digital Forensic Readiness Challenge in Organizations
   v. Trust of Audit Trails

Challenges can also be clustered and analyzed according to diverse categories for example standards, data acquisition, investigation, and reporting [28] (Figure 3.4).

Research identifies six main sources of errors in DF, as mentioned below [29]:

- Client errors.
- Wider investigative/forensic team errors.
- Practitioner errors.
- Tools/instruments errors.
- Method errors.
- Trace errors.

Machine learning contributes directly to trace errors, which occur at the investigation stage and focus on identifying and understanding digital traces such as data processing/examination, analysis, interpretation, and evaluation, and that may lead to "Missed data whereby erroneous traces may have a structure that ultimately means any tools and methods may fail to identify them (resulting in them being missed during an examination), misrepresentation: whereby traces may be parsed and

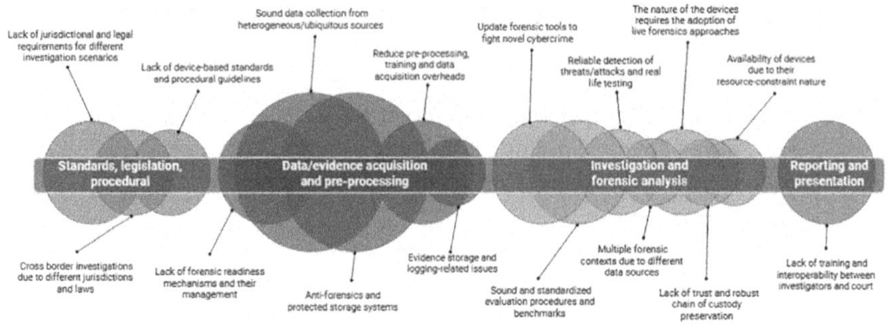

FIGURE 3.4 Challenges in DF (Source: Casino et al. 2022).

displayed incorrectly impacting any subsequent review of them by practitioners, and general misinterpretation. Trace errors have the potential to compromise the value of digital investigations" [30].

In addition to the unintended consequence of bias and error, DF processes can be complicated further by explicit misrepresentation of the facts with the intention to deceive, such as fabrication, when evidence is made up, tampering when evidence is manipulated, or omission, where evidence is not taken into account. Either of these occurrences is common, difficult to identify, and compromise the validity of any outcome.

3.5 REASONING

One of the features of intelligence is, arguably, the ability to reason. Although reasoning has been a main branch of classical logic and a field of study in many disciplines since the beginning of human history, all the knowledge contained in the field of logic cannot explain the totality of a universe of discourse, with each approach having limitations that must be acknowledged [31].

The main types of reasoning are known as inductive, abductive, and deductive, whereby induction begins with data and produces concepts; in abduction [32], relationships among the concepts are inferred to develop interrelated hypotheses; and in deduction, data is gathered to fill in the gaps and produce hypotheses. Abduction is "process of forming an explanatory hypothesis. It is the only logical operation which introduces any new idea" [32].

Experts and analysts aiming to capture truth combine all three types of reasoning [29]. Thanks to advancements in computer science, the classification of reasoning types has been extended to include diagrammatic, abstract, spatial-temporal, fuzzy, and probabilistic reasoning. It is not in the scope of this paper to discuss logic in detail; however, being reasoning the backbone of all DF capabilities, one of its cornerstones is causal inference [33], described as the ability of an algorithm to establish causal relationships among factors being analyzed. In causal inference, confounds are fact*ors that influence both the dependent and independent variable*, causing spurious associations.

Confounding factors are inherent statistical errors caused by confounding data that naturally exist in any model. But they can also be fabricated and injected into the evidence deliberately to mislead inquiry. There is no known mechanism in DF to identify, represent, and address malicious confounding, and no reliable algorithms for its identification. Expert reasoning is characterized by heuristic informal judgmental rules which guide towards plausible paths to follow and away from implausible ones [34, 35]. In solving complex problems, investigators apply heuristic expert reasoning utilizing a mix of seemingly arbitrary logical composition, so-called fuzzy causal inference:

> *In the complex domain of causal inference, the challenge has always been to extract clear insights from the often vague and imprecise data coming from real-world scenarios. These frameworks have provided robust methodologies and points of views to find the cause-and-effect relationships from both observational and experimental*

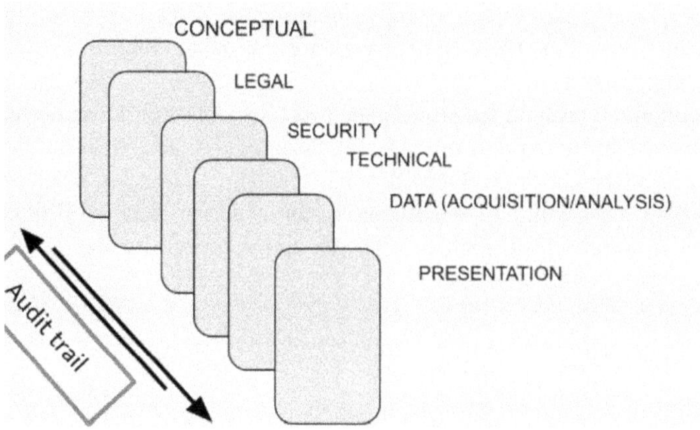

FIGURE 3.5 DF Model layers (inspired by Renaud 2021).

> *data. However, they are in trouble for dealing with the inherent fuzziness of real-life data and the subjectivity of the human decision-making* [36].

Although still a relatively new approach, heuristic reasoning can, at least in part, be automated with fuzzy logic. The automation of hybrid logical schemas and fuzzy inference in DF using AI can deliver benefits and advance machine reasoning towards more sophisticated human-like intelligence, yet it requires careful and explicit representation.

When using machine learning to carry out complex deductive reasoning involving large digital datasets, a lot can go wrong. Structured analysis approaches can help guide the adherence of DF processes to the correct protocol, evaluate the correctness of the outcomes, and generate explicit, human-readable documentary evidence. A layered framework is an example of an explicit knowledge representation that can be used to mitigate bias and error, provided it can support an audit trail [37] across the layers to ensure process integrity (Figure 3.5).

3.6 TOWARDS OPEN STANDARDS

Open-source data and information have started to become accepted as admissible evidence in mainstream judicial proceedings where it has been referred to as digital open-source information (DOSI) [38]. Open standards can be crucial for ensuring that forensic practices are consistent, interoperable, and transparent across different tools and systems. They are intended to provide a common framework to facilitate the reliable exchange of data and results between various stakeholders. Yet, in reality, as of the time of writing, open standards for digital forensics are still being tentatively drafted. The main DF standards in use remain proprietary and as such require a license to be accessed and used (see Table 4.1). A notable exception is DF XML [39], which is currently maintained by National Institute of Standards and Technology

(NIST)[2] and which is limited to representing metadata and provenance information of a subset of DF entities such as disk images; files; file system metadata; moves, adds, and changes (MAC) times; file hashes; sector hashes; transmission control protocol (TCP) flows; and hash sets.

DF XML does not provide a standardized representation of the state of the art in DF and AI at the time of writing.

Main standards in DF:

- ISO/IEC 27043:2015 – "Incident investigation principles and processes": outlines the principles and processes for conducting thorough and effective investigations of digital incidents.
- NIST Special Publication 800–101 – "Guidelines on Mobile Device Forensics": This publication provides guidelines for the forensic examination of mobile devices, including smartphones and tablets.
- NIST Special Publication 800–86 – "Guide to Integrating Forensic Techniques into Incident Response": This guide outlines how to integrate forensic techniques into incident response procedures to improve the effectiveness of investigations.
- NIST Special Publication 800–92 – "Guide to Computer Security Log Management": This publication provides guidance on the management and analysis of computer security logs, which are crucial for forensic investigations.
- ACPO (Association of Chief Police Officers) Guidelines – "Good Practice Guide for Computer-Based Evidence": These guidelines provide best practices for handling computer-based evidence to ensure that it remains intact and reliable.
- SANS Institute's "Forensic Readiness" : This guide emphasizes the importance of preparing for potential forensic investigations by implementing practices that facilitate the recovery and analysis of digital evidence.[3]
- SWGDE (Scientific Working Group on Digital Evidence) Guidelines – "Digital Evidence Guidelines": These guidelines cover various aspects of digital evidence handling and analysis, promoting consistency and reliability in forensic practices.
- Forensic Science Regulator's Code of Practice and Conduct (UK) – "Code of Practice and Conduct for Digital Forensics": This code provides standards for digital forensic practices in the UK, ensuring that forensic work is conducted to a high standard.
- **ASTM E1188–11** – Standard Practice for Collection and Preservation of Information and Physical Items by a Technical Investigator.

There are innumerable known benefits that can be derived from the adoption of open technical standards; above all, increased transparency [40], interoperability [41], encouraging innovation [42], and reducing costs associated with proprietary technologies and vendor lock-in. Organizations can avoid costly licensing fees and choose from a variety of solutions that comply with the open standard, leading to more competitive pricing [43].

Machine readability can facilitate the automation of parsing documents by computers, but to evaluate and use technical and engineering artifacts, they must be human-readable. Readability metrics of technical documentation include reading speed, text comprehension [44], and the ability to interpret, comment on, and contribute to the technical document [45].

Human readability in AI can be achieved using natural language and visualizations as well as user-friendly interfaces to query and analyze the algorithms in use. Therefore, the case for human-readable DF artifacts through the adoption of human-friendly open standards supported by explicit logical schemas for DF is implicit and very strong. A human-readable model for DF that supports explicit knowledge representation [46] can serve as the outline of a model card and is proposed here as the basis for the development of an open standard in AI for DF.

Model cards (MC) are described as "short pieces of documentation that can be leveraged by ML applications to summarize and make transparent machine learning models, as well as executable programs to test/evaluate inputted data using a model" [47].

MCs are declarative frames [48]: explicit knowledge structures written in natural language that describe the knowledge (declarative) and processes (procedures) used by the AI. In declarative modeling, models declare a set of facts that are true about the model.

3.7 MODEL CARDS IN DIGITAL FORENSICS

Capturing and making explicit the many facets of DF in a single frame may not be feasible; however, a set of nested frames can be adopted – a series of nested model cards (DF MC9 and 1, for example).

An indicative example of how model cards can be designed to represent core elements of the DF process so that it can be standardized and made transparent is illustrated in Figures 3.6 and 3.7.

The DF MC 0 (Figure 3.6) includes the representation of data types, reasoning types, and sources of errors. Using a symbolic representation (in the form of natural language or pseudocode description):

DF ModelCard_Elements (List)

- MMCID – Identifier
- MCV – Version
- DF-MCO Owner
- DF-MC Use (standalone, integrated)
- DF-MC CS (Case statement)
- DF-MC H (Hypothesis)
- DF-MC C (Classification)
- DF-MC TR (Type of reasoning)
- DF-MC B (Bias)
- DF-MC CB (Cause of Bias)

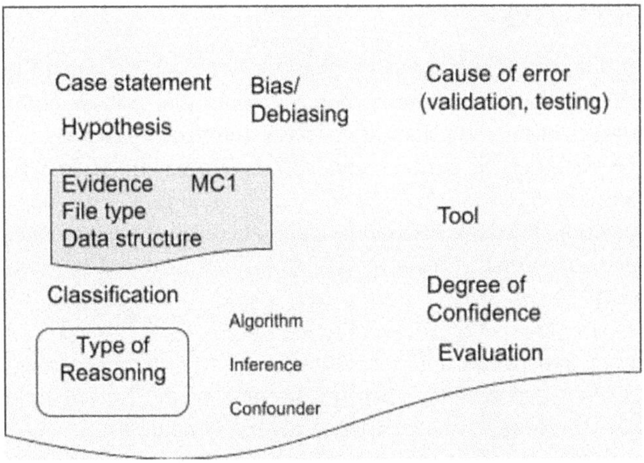

FIGURE 3.6(a) DF MC 0 – Top Level.

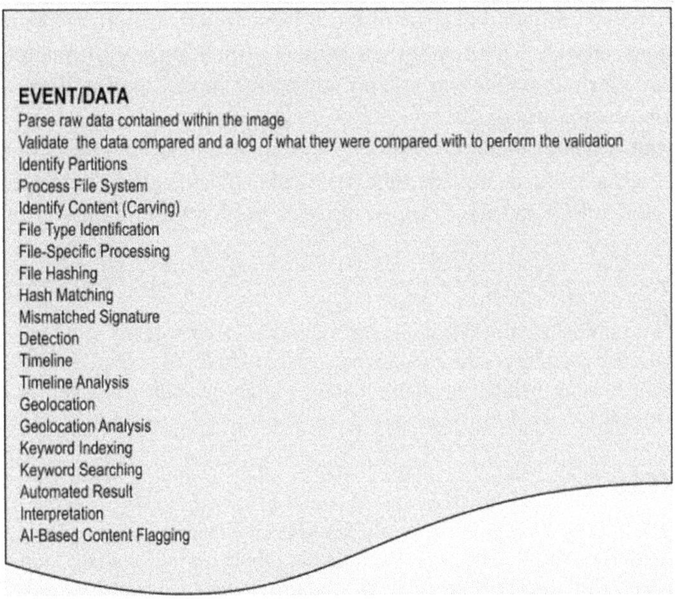

FIGURE 3.6(b) DF MC 1 Includes a specification of data and processes associated with the data; DF MC 1 = DF Data types and analytical processes (based on Heargraves 2024).

- DF-MC E (Error)
- DF-MC CE (Cause of Error)
- DF-MC Ln (Layer n)
- Others

3.8 FUTURE WORK

DF as a discipline can be advanced by machine learning, provided the formalisms adopted adequately support the multidimensionality and transdisciplinarity of the domain by understanding the criticality of its systemic complexity.

AI must be designed to address and correct known flaws in DF, rather than reinforcing them.

How can machine learning be used in DF so that it is not used to support miscarriages of justice nor to distort truth or deviate from ethical and just conclusions and decision-making?

There is no simple answer. If machine learning is to improve DF processes and make them more transparent and accountable, and to mitigate human errors rather than automate their propagation, then the fairness of the line of inquiry must be unambiguously represented, made explicit, and easily auditable.

This chapter analyzes and summarizes the state of the art, paying particular attention to the current challenges in DF and AI, providing pointers to issues that limit the reliability of ML adoption. It contributes arguments and justification towards the development of human-readable open standards and provides the outline of model cards as a schema grounded in state-of-the-art research that captures the multiplicity of dimensions provided for further discussion, validation, and refinement. Future work includes formal evaluation and proposing the model card and elements as an extension of existing standards.

About the Author: Paola Di Maio is a systems analyst, independent researcher, and a research scholar at the Institute for Global Distributed Open Research and Education (IGDORE) and the Chair of the W3C AI KR Community Group.

NOTES

1 https://www.nnlm.gov/guides/data-glossary/data-management-plan
2 https://www.loc.gov/preservation/digital/formats/fdd/fdd000611.shtml
3 https://www.sans.org/white-papers/ultimate-guide-getting-started-digital-forensics-incident-response/

REFERENCES

[1] Casey, E. (2009). *Handbook of digital forensics and investigation*. Academic Press.

[2] Cranor, L. F. (2008). A framework for reasoning about the human in the loop. *UPSEC'08: Proceedings of the 1st conference on usability, psychology, and security* (pp. 1–15). https://dl.acm.org/doi/10.5555/1387649.1387650

[3] Garfinkel, S. (2010). Digital forensics research: The next 10 years. *Digital Invest*, 7, S64–73.

[4] Himmelsbach, T., Mou, Y., Decker, S., & Heinzl, A. (2023). Towards federated machine learning and distributed ledger technology-based data monetization. In *VLDB workshops*. ceur-ws.org

[5] Jones, G. M., & Winster, S. G. (2022). An insight into digital forensics: History, frameworks, types and tools. *Cyber Security and Digital Forensics*, 105–125.

[6] Ahmed, R., & Dharaskar, R. V. (2008). Mobile forensics: An overview, tools, future trends and challenges from law enforcement perspective." In *6th international conference on e-governance, ICEG, emerging technologies in e-government, m-government* (pp. 312–323). http://www.iceg.net/2008/. 17-19 April 2009 3rd IEEE/ACM

[7] Meghanathan, N., Allam, S. R., & Moore, L. A. (2010). Tools and techniques for network forensics. *arXiv preprint arXiv:1004.0570*.
[8] Murdoch, D., Xu, J., Liu, S., Shen, Y., & Yu, C. (2018). Big data forensics: Challenges and opportunities. *IEEE Cloud Computing*, *5*(1), 50–57.
[9] Noyes, J., & Keegan, M. (2019). Big data analytics in digital forensics. In A. Parisi & M. Piattini (Eds.), *Digital forensics and cyber crime* (pp. 289–305). Springer International Publishing.
[10] Himmelsbach, M., & Themis, M. (2020). The role of digital forensics in data science. In *International conference on cybercrime (ICCC)* (pp. 122–129). IEEE; Garfinkel, S. (2010). *Digital forensics science: The best practices illustrated*. Addison-Wesley Professional.
[11] Dunsin, D., Ghanem, M. C., Ouazzane, K., & Vassilev, V. (2024). A comprehensive analysis of the role of artificial intelligence and machine learning in modern digital forensics and incident response. *Forensic Science International: Digital Investigation*, *48*, 301675.
[12] Al-Jadir, I., Wong, K. W., Fung, C. C., & Xie, H. (2018). October. Enhancing digital forensic analysis using memetic algorithm feature selection method for document clustering. In *2018 IEEE international conference on systems, man, and cybernetics (SMC)* (pp. 3673–3678). IEEE.
[13] Dimitriadis, A., Lontzetidis, E., Kulvatunyou, B., Ivezic, N., Gritzalis, D., & Mavridis, I. (2022). Fronesis: Digital forensics-based early detection of ongoing cyber-attacks. *IEEE Access*, *11*, 728–743.
[14] Zhang, Q., Yang, L. T., Chen, Z., & Li, P. (2018). A survey on deep learning for big data. *Information Fusion*, *42*, 146–157.
[15] Faehndrich, J., Honekamp, W., Povalej, R., Rittelmeier, H., Berner, S., & Labudde, D. (2023). Digital forensics and strong AI: A structured literature review. *Forensic Science International: Digital Investigation*, *46*, 301617.
[16] Cole, S. A. (2014). *Forensic science and miscarriages of justice*. Springer Science+Business Media.
[17] Etter, B. (2013). The contribution of forensic science to miscarriage of justice cases. *Australian Journal of Forensic Sciences*, *45*(4), 368–380.
[18] Sunde, N., & Dror, I. E. (2019). Cognitive and human factors in digital forensics: Problems, challenges, and the way forward. *Digital Investigation*, *29*, 101–108.
[19] Dror, I. E. (2020). Cognitive and human factors in expert decision making: Six fallacies and the eight sources of bias. *Analytical Chemistry*, *92*(12), 7998–8004.
[20] Moser, S. (2013). Confirmation bias: The pitfall of forensic science. *Themis: Research Journal of Justice Studies and Forensic Science*, *1*, Article 7.
[21] Renaud, K., Bongiovanni, I., Wilford, S., & Irons, A. (2021). PRECEPT-4-Justice: A bias-neutralising framework for digital forensics investigations. *Science & Justice*, *61*(5), 477–492.
[22] Jinad, R., Gupta, K., Simsek, E., & Zhou, B. (2024). Bias and fairness in software and automation tools in digital forensics. *Journal of Surveillance, Security and Safety*, *5*, 19–35.
[23] Sunde, N., & Dror, I. E. (2019). Cognitive and human factors in digital forensics: Problems, challenges, and the way forward. *Digital Investigation*, *29*, 101–108.
[24] Karie, N. M., & Venter, H. S. (2015). Taxonomy of challenges for digital forensics. *Journal of Forensic Sciences*, *60*(4), 885–893.
[25] Mijwil, M., & Aljanabi, M. (2023). Towards artificial intelligence-based cybersecurity: The practices and ChatGPT generated ways to combat cybercrime. *Iraqi Journal For Computer Science and Mathematics*, *4*(1), 65–70.
[26] Dempsey, R. P., Brunet, J. R., & Dubljević, V. (2023). Exploring and understanding law enforcement's relationship with technology: A qualitative interview study of police officers in North Carolina. *Applied Sciences*, *13*(6), 3887.
[27] Swofford, H., & Champod, C. (2022). Probabilistic reporting and algorithms in forensic science: Stakeholder perspectives within the American criminal justice system. *Forensic Science International: Synergy*, *4*, 100220.

[28] Casino, F., Dasaklis, T. K., Spathoulas, G. P., Anagnostopoulos, M., Ghosal, A., Borocz, I., Solanas, A., Conti, M., & Patsakis, C. (2022). Research trends, challenges, and emerging topics in digital forensics: A review of reviews. *IEEE Access*, *10*, 25464–25493

[29] Åsvoll, H. (2013). Abduction, deduction and induction: Can these concepts be used for an understanding of methodological processes in interpretative case studies? *International Journal of Qualitative Studies in Education*, *27*(3), 289–307. https://doi.org/10.1080/09518398.2012.759296.

[30] Horsman, G. (2024). Sources of error in digital forensics. *Forensic Science International: Digital Investigation*, *48*, 301693.

[31] Yanofsky, N. S. (2016). *The outer limits of reason: What science, mathematics, and logic cannot tell us*. MIT Press.

[32] Peirce, C. S. (1992). *The essential Peirce, volume 2: Selected philosophical writings (1893–1913)* (Vol. 2). Indiana University Press.

[33] Yao, L., Chu, Z., Li, S., Li, Y., Gao, J., & Zhang, A. (2021). A survey on causal inference. *ACM Transactions on Knowledge Discovery from Data (TKDD)*, *15*(5), 1–46.e

[34] Polya, G. (1945). *How to solve it*. Princeton, NJ: Princeton University Press.

[35] Lenat, D. B. (1982). The nature of heuristics. *Artificial Intelligence*, *19*(2), 189–249.

[36] Saki, A., & Faghihi, U. (2024). Integrating fuzzy logic with causal inference: Enhancing the pearl and Neyman-Rubin methodologies. *arXiv preprint arXiv:2406.13731*.

[37] Surden, H. (2019). Machine learning and the law. *Harvard Journal on Legislation*, *56*(1), 25–65.

[38] Gillett, M., & Fan, W. (2023). Expert evidence and digital open source information: Bringing online evidence to the courtroom. *Journal of International Criminal Justice*, *21*(4), 661–693.

[39] Garfinkel, S. (2012). Digital forensics XML and the DFXML toolset. *Digital Investigation*, *8*(3–4), 161–174.

[40] Russell, A. L. (2014). *Open standards and the digital age*. Cambridge University Press.

[41] Berners-Lee, T. (2005). *Weaving the web: The original design and ultimate destiny of the world wide web by its inventor*. San Francisco: Harper.

[42] Mowery, D. C., & Rosenberg, N. (1991). *Technology and the pursuit of economic growth*. Cambridge University Press.

[43] David, P. A., & Foray, D. (2003). Economic fundamentals of the knowledge society. *Policy Futures in Education*, *1*(1), 20–49.

[44] Kriukova, A. (2022). *Measuring readability of technical texts*. Institute of Formal and Applied Linguistics (Master thesis).

[45] Oliveira, D., Bruno, R., Madeiral, F., & Castor, F. (2020, September). Evaluating code readability and legibility: An examination of human-centric studies. In *2020 IEEE international conference on software maintenance and evolution (ICSME)* (pp. 348–359). IEEE.

[46] Hargreaves, C., Nelson, A., & Casey, E. (2024). An abstract model for digital forensic analysis tools-A foundation for systematic error mitigation analysis. *Forensic Science International: Digital Investigation*, *48*,

[47] Mitchell, M., Wu, S., Zaldivar, A., Barnes, P., Vasserman, L., Hutchinson, B., Spitzer, E., Raji, I. D., & Gebru, T. (2019, January). Model cards for model reporting. In *Proceedings of the conference on fairness, accountability, and transparency* (pp. 220–229). ACM. Org.

[48] Winograd, T. (1975). Frame representations and the declarative/procedural controversy. In *Representation and understanding* (pp. 185–210). Morgan Kaufmann.

4 Identifying Patterns through Integration of Natural Language Processing and Semantic Web Technologies

Tatiana Avdeenko and Artem Golubnichiy

4.1 INTRODUCTION

From the very beginning of the emergence of artificial intelligence, natural language processing (NLP) was undoubtedly declared to be the most important function of a computer program claiming to be intelligent. Indeed, Alan Turing, in 1950, in his famous article [1], proposed NLP as the first of four capabilities of a computer program for passing the Turing test. According to [1], in addition to natural language processing tools, an intelligent program should include knowledge representation tools, with the help of which the computer can record in memory what it learns or reads, tools for automatically generating logical conclusions, providing the ability to use stored information to find answers to questions and draw new conclusions, as well as machine learning tools that allow learning and thus adapting to new (not programmed) situations. Natural language processing has become especially interesting and popular in recent decades in light of the success of large language models.

Natural language processing is based on two concepts: "natural language" and "processing." Natural language is best understood as a language that develops with human culture, is used for communication between people, and evolves over time. At the same time, the second term "processing" may be understood not so unambiguously and largely depends on the authors and the tasks they solve. Thus, Dingli Chen in [2] identifies two main components of natural language processing: natural language understanding (NLU) and natural language generation (NLG). From the standpoint of the challenges to be solved in the field of natural language processing, he identifies two important aspects: technical challenges and linguistic challenges. At the same time, the author attributes the creation of language corpora to technical challenges, also pointing out the limitations of pure corpus linguistics in terms of the lack of historical and cultural context within corpora. The main, in the understanding

DOI: 10.1201/9781003512820-4

of the author of the work [2], most critical, are linguistic challenges, which in turn face the following problems:

- Diversity of languages and the resulting lack of a standardized approach for their processing by machines;
- Ambivalence of language, leading to ambiguity;
- Robustness of language, expressed in the need for stable performance when processing different accents and speaking styles;
- Knowledge dependence, which is both a linguistic and technical challenge, expressed from practical knowledge;
- Contextual understanding, based on the environment and context.

The above problems are solved in phonetics, semantics, and pragmatics. At the same time, automatic text processing can largely depend on the subject area. In turn, the process of automatic text processing is closely related to the concept of patterns. The concept of pattern in the processing of language structures is fundamentally important. The patterns themselves have diverse natures and different levels of abstraction. A generalization of the degree of abstraction of patterns is shown in Figure 4.1.

At the lowest level of abstraction in text processing, the first pattern format is character patterns, a typical example of which are regular expressions. Regular expressions (Regex) are a powerful tool for working with text data. They are strings that describe a search pattern that can be used for various text processing tasks. Regular expressions allow operations such as searching, replacing, checking, and extracting substrings. As patterns, regular expressions can be used to automatically extract and process text data, which is especially useful in the following cases:

1. *Data Extraction*. Regular expressions can be used to extract specific parts of text from large amounts of data, such as log files, HTML pages, or text files;
2. *Pattern Validation*. Regular expressions can be used to check whether strings follow a certain format. For example, validating user input such as an email address, phone number, or zip code;
3. *Replacement and Formatting*. Regular expressions can be used to replace or format text. For example, replacing all URLs in a text with hyperlinks;
4. *Data Parsing*. Regular expressions are used to parse text data, i.e., to separate text into its constituent parts according to certain rules. For example, splitting an event log into individual entries.

All the designated areas of application of regular expression patterns have practical significance for technical analysis. At the same time, regular expressions are a practical example of implementing several types of patterns at once, allowing you to extract characters and phrases. Regular expressions are also implemented in other pattern searches but require more complex settings.

The present paper proposes an approach to natural language processing using semantic web technologies. Unlike existing studies, the focus is on developing an algorithm that integrates NLP and semantic web methods at the initial stage of data processing, which provides deeper analysis and interpretation of text information.

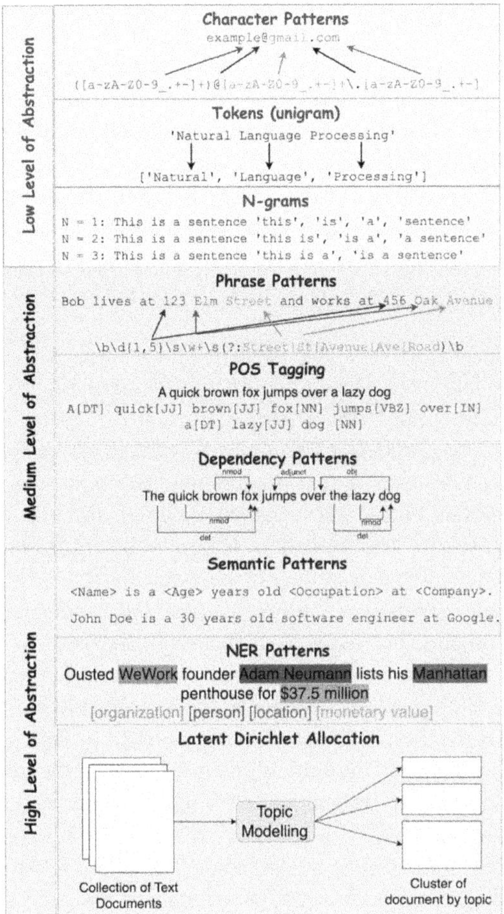

FIGURE 4.1 Patterns by Abstraction Level.

The paper has the following structure. Section 4.2 provides an overview of NLP methods and semantic web technologies used for text analysis. Section 4.3 considers semantic web technologies and ontologies in the context of pattern identification problems and describes the methods of their application. Section 4.4 delves into the description of machine learning methods and explores their applicability in identifying patterns within data analysis tasks. Section 4.5 examines the integration of machine learning techniques with semantic web technologies, emphasizing how the fusion of diverse data sources enhances accuracy in problem-solving. Section 4.6 outlines the proposed algorithm designed for pattern identification tasks, providing detailed examples of its implementation in named entity recognition (NER). Finally, Section 4.7 presents the results of applying the optimized NER algorithm, alongside an analysis of existing limitations and challenges in utilizing semantic web technologies and NLP for text analysis.

4.2 FUNDAMENTALS OF NATURAL LANGUAGE PROCESSING

Natural language processing (NLP) is a discipline dedicated to studying the capabilities and techniques for processing natural language through computational technologies. It encompasses not only the interpretation and transformation of natural language text but also the generation of meaningful text and the extraction of valuable, often implicit, information embedded within it. Traditionally, NLP is divided into the following subfields: morphological analysis, syntactic analysis, and semantic analysis.

4.2.1 Morphological Analysis

Morphological analysis involves the process of breaking texts into the smallest meaningful units of language. Depending on the task, such units are created through lemmatization, stemming, or POS-tagging.

Lemmatization is the process of reducing a word to its base form. This procedure is essential because, during text processing, each new sequence of characters is perceived as a unique entity. For instance, singular and plural forms of a word or different verb tenses will be treated as separate objects for processing. Such an approach can complicate the extraction of statistical characteristics from the text [3]. Therefore, in practice, words like "authentication" and "authenticated" are reduced to the base form "authenticate" to avoid duplicating concepts.

The process of reducing a word to its base form, as described above, is a best practice and often yields good results. However, in real-world scenarios, less precise but sufficiently effective methods can also be applied [4]. One such less resource-intensive method is stemming. Stemming involves truncating a word to its root form. For example, with the words "authentication" and "authenticated," stemming would result in the root "authent."

A more advanced approach to morphological analysis is part-of-speech (POS) tagging. POS tagging is a process where each word is assigned a grammatical category. For languages where the order of words belonging to different grammatical categories is strictly defined by rules, this approach facilitates identifying semantic relationships between words and significantly simplifies the extraction of useful information [5]. For example, in our case, the word "authenticate" would be tagged as a verb, while "authentication" would be tagged as a noun.

In Figure 4.2, a code snippet is provided demonstrating the use of the spaCy library for morphological analysis.

Morphological analysis serves to normalize words within a text by reducing them to their base forms, or lemmas, which facilitates more efficient and accurate analysis and interpretation. For instance, the word "attempting" is reduced to its base form "attempt," and "detected" becomes "detect." This normalization unifies different word forms, enabling a more systematic and precise analysis. Additionally, part-of-speech (POS) tagging assigns each word a specific grammatical category, which aids in understanding the syntactic structure of sentences. For example, the word "attacker" is identified as a noun (NOUN), while "bypass" is tagged as a verb (VERB). The application of POS tagging enhances the identification of semantic relationships between words and helps establish contextual understanding. In domains

```
import spacy

nlp = spacy.load("en_core_web_sm")
doc = nlp("The attacker was attempting to bypass the firewall,
but the security system detected the intrusion.")

for token in doc:
print(f"Word: {token.text}, Lemma: {token.lemma_}, POS: {token.pos_}")


Word: The, Lemma: the, POS: DET
Word: attacker, Lemma: attacker, POS: NOUN
Word: was, Lemma: be, POS: AUX
Word: attempting, Lemma: attempt, POS: VERB
Word: to, Lemma: to, POS: PART
Word: bypass, Lemma: bypass, POS: VERB
Word: the, Lemma: the, POS: DET
Word: firewall, Lemma: firewall, POS: NOUN
Word: ,, Lemma: ,, POS: PUNCT
Word: but, Lemma: but, POS: CCONJ
Word: the, Lemma: the, POS: DET
Word: security, Lemma: security, POS: NOUN
Word: system, Lemma: system, POS: NOUN
Word: detected, Lemma: detect, POS: VERB
Word: the, Lemma: the, POS: DET
Word: intrusion, Lemma: intrusion, POS: NOUN
Word: ., Lemma: ., POS: PUNCT
```

FIGURE 4.2 An Example of Using the spaCy Library for Morphological Analysis.

such as digital forensics, this process is particularly valuable for identifying the presence and roles of critical terms within a text, such as "bypass," "detect," and similar keywords, along with their positional relationships. This deeper analysis supports more accurate interpretation and decision-making in specialized fields.

4.2.2 Syntactic Analysis

The primary objective of syntactic analysis is to examine the sentence structure and determine the grammatical relationships between individual words. This is achieved through dependency tree analysis and the creation of a syntactic tree. Dependency tree analysis constructs a graph where nodes correspond to words and edges represent the grammatical relationships between them. Unlike typical graph structures, these edges do not carry numerical weights. Instead, the connections are annotated with labels that specify the type and nature of relationships between words. Despite the absence of numerical weights, these annotations effectively render the graph "weighted," as they encode crucial syntactic information. This approach provides a clear and organized representation of the text's structure and its components, facilitating deeper analysis and understanding [6].

The construction of a syntactic tree allows for the creation of a more complex hierarchical structure. Syntactic trees help establish hierarchies between groups of words, sentences, and parts of speech. This approach not only aids in better understanding the syntactic structure of the text but also facilitates the interpretation of its semantics. In [7], three generative lexicalized models are presented that, based on statistical analysis and probabilistic methods, enable more accurate analysis of complex sentences and improve the quality of syntactic text analysis. The code example below demonstrates syntactic analysis of text using the spaCy library.

The output of the code confirms that syntactic analysis identifies grammatical relationships between words. During the analysis, relationships such as "head-dependent" are established. For example, in the sentence from the example, it was determined that "attacker" is the subject of the verb "tried," which, in turn, serves as the root of the entire sentence. In the context of implementation in the spaCy library, it is marked with the label ROOT. The word "bypass" acts as a dependent action (xcomp) of "attempting," and "firewall" is a direct object (dobj) of "bypass." The conjunction "but" (cc) connects two actions, "attempting" and "detected," where "system" is the subject (nsubj) of "detected," and "intrusion" is its direct object (dobj). This analysis helps us understand how different elements of the sentence interact to form its semantic and grammatical structure.

4.2.3 Semantic Analysis

The main goal of semantic analysis is to uncover the meaning of words and sentences. Depending on the level of abstraction, this task can fall into one of three categories: named entity recognition, semantic role labeling, and anaphora resolution.

```
import spacy

nlp = spacy.load("en_core_web_sm")
doc = nlp("The attacker was attempting to bypass the firewall, but
the security system detected the intrusion.")

for token in doc:
print(f"Word: {token.text}, Dependency: {token.dep_},
Head: {token.head.text}")

Word: The, Dependency: det, Head: attacker
Word: attacker, Dependency: nsubj, Head: attempting
Word: was, Dependency: aux, Head: attempting
Word: attempting, Dependency: ROOT, Head: attempting
Word: to, Dependency: aux, Head: bypass
Word: bypass, Dependency: xcomp, Head: attempting
Word: the, Dependency: det, Head: firewall
Word: firewall, Dependency: dobj, Head: bypass
Word: ,, Dependency: punct, Head: attempting
Word: but, Dependency: cc, Head: attempting
Word: the, Dependency: det, Head: system
Word: security, Dependency: compound, Head: system
Word: system, Dependency: nsubj, Head: detected
Word: detected, Dependency: conj, Head: attempting
Word: the, Dependency: det, Head: intrusion
Word: intrusion, Dependency: dobj, Head: detected
Word: ., Dependency: punct, Head: detected
```

FIGURE 4.3 An Example of Using the spaCy Library for Syntactic Analysis.

Named entity recognition (NER) is the process of identifying and classifying named entities, where each word or phrase is assigned a label indicating its entity type. Such labels can include categories like people, places, and dates. Extracting named entities enables the retrieval of meaningful information from text, which is particularly useful for tasks like information retrieval and the analysis of large datasets. Approaches to recognizing and classifying named entities, as presented in [8], explore this method from the perspective of semantic analysis.

A higher level of abstraction compared to NER is Semantic Role Labeling (SRL). This approach identifies the roles of words within a sentence, such as the active entity (agent) that performs an action, the entity (patient) upon which the action is executed, as well as additional objects and contextual elements that define the interaction. Building upon earlier methods, this approach provides a deeper and more detailed understanding of both the structural and semantic aspects of a sentence, enabling more accurate analysis and interpretation [9]. Anaphora resolution involves determining the referents of pronouns and other similar expressions in a text. This process establishes connections between different parts of the text and, as shown in [10], enhances the understanding of text semantics. Figure 4.4 provides an example of applying NER to a text in the field of computer security, along with the results of entity extraction using the spaCy library.

The task of NER can be addressed by leveraging large language models, such as those based on transformers. These models are widely applicable in the field of NLP. An example implementation of this task using a model for automatic token classification from the transformers module is provided in Figure 4.5.

A detailed comparison of the results obtained using the spaCy library and Transformer highlights the differences in their approaches and capabilities for named entity recognition (NER). The spaCy library identifies entire entities without fragmenting their meaning, which is particularly valuable for terms like "Microsoft Corporation" or "FBI." Additionally, it employs more specific labels, such as "GPE" (Geopolitical Entity), to classify geopolitical objects, enhancing the precision of its categorization. However, spaCy also has notable limitations. For example, it struggled to recognize

```
import spacy

nlp = spacy.load("en_core_web_sm")

doc = nlp("On July 15, 2023, a major cyber attack targeted Microsoft
Corporation, compromising the data of over 500,000 users. The attack
was traced back to a hacker group operating out of Moscow, Russia.
The FBI and Interpol are currently investigating the incident")

for ent in doc.ents:
print(f"Entity: {ent.text}, Label: {ent.label_}")

Entity: July 15, 2023, Label: DATE
Entity: Microsoft Corporation, Label: ORG
Entity: over 500,000, Label: CARDINAL
Entity: Moscow, Label: GPE
Entity: Russia, Label: GPE
Entity: FBI, Label: ORG
Entity: Interpol, Label: ORG
```

FIGURE 4.4 An Example of Using the spaCy Library for NER.

```
from transformers import AutoTokenizer, AutoModelForTokenClassification, pipeline

tokenizer = AutoTokenizer.from_pretrained("dslim/bert-base-NER")
model = AutoModelForTokenClassification.from_pretrained("dslim/bert-base-NER")

nlp = pipeline("ner", model=model, tokenizer=tokenizer)
sentence = "On July 15, 2023, a major cyber attack targeted Microsoft Corporation,
compromising the data of over 500,000 users. The attack was traced back to a hacker
group operating out of Moscow, Russia. The FBI and Interpol are currently
investigating the incident"

ner_results = nlp(sentence)

for result in ner_results:
print(f"Token: {result['word']}, Entity: {result['entity']},
Score: {result['score']:.4f}")

Token: Microsoft, Entity: B-ORG, Score: 0.9994
Token: Corporation, Entity: I-ORG, Score: 0.9993
Token: Moscow, Entity: B-LOC, Score: 0.9997
Token: Russia, Entity: B-LOC, Score: 0.9998
Token: FBI, Entity: B-ORG, Score: 0.9992
Token: Inter, Entity: B-ORG, Score: 0.9996
Token: ##pol, Entity: I-ORG, Score: 0.9992
```

FIGURE 4.5 An Example of Using a Model for Automatic Token Classification from the Transformers Module.

certain complex or compound entities, such as "Interpol," which the Transformer model successfully identified. This distinction underscores the Transformer model's superior ability to handle more intricate linguistic constructs, offering broader applicability in complex NER tasks. The Transformer-based model, on the other hand, operates at the level of tokens and sub-tokens, which sometimes results in splitting entities into smaller parts, not always accurately. However, this approach allows Transformers to handle complex and rare entities more flexibly.

The differences described above are fully explained by the architecture of the models. Transformers, with their ability to consider context at a deeper level, are more effective at recognizing complex and unusual entities. However, their tendency to break words into sub-tokens can lead to fragmented prediction results. In contrast, spaCy, being a more traditional model for natural language processing, is better suited for tasks where it is important to accurately and holistically recognize known and well-defined entities. As a result, the choice of model for a particular task will depend on the requirements for accuracy and the ability to handle complex or composite entities in the text. To assess the completeness and accuracy of entity recognition, it is advisable to conduct synthetic tests or natural tests on a large amount of corpus data with pre-labeled entities. The results of the study of natural tests are beyond the scope of this study, so we will limit ourselves to just a small example on artificially generated text in comparison of the spaCy model and the Flair model for the English language. The source text for testing the NER models is provided in Appendix A.

This study compared two popular named entity recognition models, spaCy and Flair, on a set of generated digital forensics text. The main objective was to evaluate the performance of these models in identifying different entity types, such

Classification Report for spaCy:

	precision	recall	f1-score	support
LOC	0.71	0.71	0.71	7
MISC	0.50	0.08	0.14	12
ORG	0.91	0.78	0.84	27
Other	0.00	0.00	0.00	4
accuracy			0.54	50
macro avg	0.53	0.39	0.42	50
weighted avg	0.71	0.54	0.59	50

Classification Report for Flair:

	precision	recall	f1-score	support
LOC	0.78	1.00	0.88	7
MISC	0.50	0.25	0.33	12
ORG	0.96	0.89	0.92	27
Other	0.00	0.00	0.00	4
accuracy			0.68	50
macro avg	0.56	0.53	0.53	50
weighted avg	0.75	0.68	0.70	50

FIGURE 4.6 Results of the Named Entity Classification Report by the spaCy and Flair Models.

as organizations (ORG), locations (LOC), and other categories (MISC), against a pre-prepared gold standard. For the experiment, the text and the gold standard were formatted in the same way, including converting all characters to lowercase to avoid comparison errors. Entities in the gold standard were grouped into their respective categories, and each model underwent entity normalization, which included removing articles and matching entity types to gold standard categories. The text was processed by both models, and the entities they predicted were compared against the reference data. To quantify the performance of the models, accuracy, precision, recall, and F1 scores were used to measure how effectively each model recognized entities and classified them correctly. The results are shown in Figure 4.6.

The obtained results demonstrate that the Flair model shows higher precision and recall rates, especially in recognizing entities in the LOC and ORG categories. However, both models have difficulties in correctly identifying entities in the MISC category.

4.3 SEMANTIC WEB AND ONTOLOGIES IN PATTERN IDENTIFICATION TASKS

The problems of semantic role analysis increasingly reveal the relationship between text processing and semantics. Ontologies and semantic web technologies serve as robust tools for addressing entity identification challenges by offering structured data representation and enabling the creation of intricate relationships between various elements. Key components of the semantic web and ontology technologies include RDF, OWL, and SPARQL, each contributing to efficient pattern identification through the modeling, description, and integration of data from diverse sources. RDF is primarily used for data modeling and integration. Its foundation on subject-predicate-object triples ensures a standardized approach, making it particularly suitable for integrating data from multiple sources. For instance, in the domain of computer security, RDF can unify datasets by aggregating data from scientific publications in digital forensics, research reports, and analytical findings. A prominent example of RDF's capabilities is the Linked Open Data project, which uses RDF to connect and integrate open data across the web [11].

Web Ontology Language (OWL) is specifically designed for creating and managing ontologies in the semantic web, enabling the definition of classes, properties, and relationships within a domain. OWL's expressive capabilities make it a valuable tool for building detailed and domain-specific ontological structures, supporting advanced reasoning and inference [12]. These technologies collectively enable effective pattern identification by uncovering recurring structures and relationships within data, making them indispensable for integrating and analyzing information across complex domains. In this context, an ontology refers to a computer ontology, which is a formal description of knowledge in a specific domain. OWL employs classes, properties, and relationships between them to create such descriptions, aligning this approach closely with object-oriented modeling. Developed by the W3C consortium, OWL is part of the semantic web and aims to improve data interoperability on the Internet.

In the context of pattern identification, two main directions can be highlighted: ontology creation and inference.

1. Creating ontologies. OWL is used to create ontologies that formally describe classes, properties, and relationships between entities. Ontologies help formalize domain knowledge, making it easier to discover patterns and relationships.
2. Inference. OWL supports inference, allowing new relationships and patterns to be automatically discovered based on specified rules and constraints, which is useful in areas where it is necessary to make inferences from large amounts of data [13].

A typical example of OWL application is the field of bioinformatics, where OWL is used to create ontologies. Thus, Gene Ontology helps identify connections and patterns in genetic data [14, 15]. Semantic web technologies are also actively used to address cybersecurity issues. In [16], the issue of ensuring cybersecurity through the creation of a relevant thematic knowledge base and its applications in practice is considered. SPARQL (SPARQL Protocol and RDF Query Language) is a query language designed to extract and manipulate data represented in the RDF (Resource Description Framework) format. SPARQL has been standardized by the W3C (World Wide Web Consortium) and is a key component of Semantic Web technology. SPARQL allows users to specify complex queries to data stored in RDF graphs and receive answers in various formats, such as JSON, XML, and CSV [17].

SPARQL occupies a special place in working with pattern identification, namely, in terms of organizing and accessing data. The main options for working with SPARQL are given below:

1. Queries to RDF data. SPARQL allows complex queries to be performed on data stored in an RDF format, which helps identify patterns and relationships. Queries can be constructed in such a way as to retrieve specific data or identify regularities. SPARQL is actively used for building knowledge organization systems. For example, in [18], options for data aggregation using graph-based knowledge are analyzed. This task is addressed in the context of anomaly detection and improving prediction accuracy, as well as for the needs of security analysts. Regarding the ability to integrate data from various sources, SPARQL is largely similar to OWL, enabling the extraction and integration of data from open sources such as DBpedia and Wikidata, which is highly effective for building large sets of linked data [19].
2. Creating new data. SPARQL not only works with existing data in its "as-is" format but also allows the creation of new RDF graphs from existing ones, which is also useful for pattern analysis tasks. This approach provides broad opportunities, including for multimedia data, as demonstrated in [20].

4.4 MACHINE LEARNING METHODS FOR PATTERN IDENTIFICATION

Machine learning methods, encompassing a wide variety of approaches, enable the automation of pattern identification processes both in small natural language texts and in large datasets. In machine learning models, such patterns serve as recurring structures, the detection of which allows for the construction of various types of dependencies. Each machine learning method, however, warrants separate consideration.

4.4.1 Classification Models

Classification models implementing supervised learning methods are designed to predict categories or labels based on input data. These models are trained on labeled datasets, where each set of input features is associated with a known label. Based on this experience, they can be used to identify and classify patterns across various types of data. Logistic regression is used for binary classification and helps identify patterns that indicate the presence or absence of a threat. For example, identifying phishing messages based on the content analysis of emails [21]. Decision trees build a tree-like structure that is used to classify data based on attribute values. For example, classifying network attacks by type (e.g., DDoS attack and SQL injection) based on network traffic characteristics [22].

Random forest consists of many decision trees that vote for the final class, which in turn improves the accuracy and robustness of the model with a large number of participants in this process. For example, identifying potential vulnerabilities in software based on the analysis of system logs and error reports [21]. The support vector machine (SVM) finds a hyperplane that maximally separates classes in a multidimensional feature space. For example, classifying malware based on its binary code and behavioral characteristics [22].

4.4.2 Clustering Models

Clustering models, which implement unsupervised learning methods, group data based on their similarities, allowing them to identify natural patterns and structures without the need for pre-known labels. Unlike classification models, which require labeled data for training, clustering models work with unlabeled data, independently identifying internal structures and patterns in the data [23]. K-means divides data into K clusters, minimizing the intra-cluster distance. In the context of cybersecurity, this method can be used to classify network traffic by activity types, helping, for example, to detect suspicious behavior that may indicate potential threats.

Hierarchical clustering creates a tree of clusters (dendrogram) by merging or dividing clusters at different levels of the hierarchy. In cybersecurity, this method can be used to analyze malware, grouping it by similar characteristics, which helps in classifying new threats based on their behavior or source code. density-based spatial clustering of applications with noise (DBSCAN) determines clusters based on data density and ignores noise. It can be used, for example, to detect anomalies in transactional data.

4.4.3 Time Series Analysis

Time series represent a unique type of data because patterns within them are structured around time. Unlike other data types, time series can exhibit explicit cyclical structures and follow rules that are uncommon in other predictive models [24]. Time series are actively used in cybersecurity for monitoring and forecasting threats, as events related to failures and DDoS attacks often have characteristic temporal features. There are numerous methods for analyzing time series, each with varying computational and temporal complexity.

Currently, the most popular method involves the use of generative adversarial networks (GANs). These networks consist of a generator and a discriminator that compete with each other to generate new data. More conventional and less resource-intensive technologies in this field are neural networks such as RNNs and LSTMs. RNNs are based on processing sequential data, taking into account information from previous states, such as identifying patterns of anomalies. LSTMs, as an extension of RNNs, enable the effective analysis of longer-term dependencies by addressing the vanishing gradient problem.

4.4.4 Examples of Using Machine Learning Methods to Identify Patterns

Some examples of cybersecurity problems solved using pattern detection are summarized in Table 4.1.

As seen in the table presented above, the most common task solved using machine learning methods is classification. Deep neural networks and classical methods are

TABLE 4.1
Application of Machine Learning Methods in Cybersecurity Tasks

Problem solved	Technologies
Detecting anomalies in network traffic	Classification models and machine learning methods
Malware detection	Clustering and time series analysis
Prediction of cyberattacks	Classification models and deep neural networks (CNN)
Generation of phishing attacks	Generative adversarial networks (GAN)
Automatic threat classification	Support vector machine (SVM)
Software vulnerability analysis	Logistic regression and decision trees
Detection and mitigation of DDoS attacks	Random forest and deep neural networks
Log analysis for intrusion detection	Hierarchical clustering and DBSCAN
Insider threat detection	Clustering and time series analysis
Data leak detection and prevention	Deep neural networks and temporal convolutional networks (TCN)
Facial recognition in access control systems	Deep neural networks (CNN)
Detection of forgery and counterfeiting of digital documents	Generative adversarial networks (GAN)
Network attack classification	Support vector machine (SVM) and random forest

used to address classification problems. For the most part, these methods are versatile and suitable for solving various tasks, as evidenced by the recurring technologies employed for different objectives. Among the more specific technologies, it is worth highlighting GANs, which are used for creating and preventing phishing attacks, as well as for detecting forgeries in digital documents.

4.5 INTEGRATION OF PATTERN RECOGNITION WITH THE SEMANTIC WEB AND DATA FUSION FROM HETEROGENEOUS SOURCES

Currently, the use of ontologies for more efficient pattern detection has not been fully explored. However, even at this stage of research, it is possible to identify and analyze the main areas and directions where semantic web technologies and ontologies can significantly enhance text analysis tasks.

4.5.1 Use of Ontologies

Ontologies represent formalized knowledge models, typically based on an object data model. A distinctive feature of ontologies is their clear structure and semantic context. This feature enables the linking of concepts, thereby improving various NLP tasks, including pattern discovery for text annotation and interpretation. Enhanced text interpretation is achieved by providing contextual information. For instance, using resources like DBpedia, it becomes possible to extract context, which, in turn, can unambiguously annotate text with named entities [25]. A related process is the reduction of ambiguity. Ontologies help address the problem of polysemy by providing extended context for accurately understanding the semantic role of terms. For example, if the word "apple" appears in a text, depending on the domain defined by the ontology, this term may refer either to the IT company or to the fruit.

4.5.2 Data Integration from Various Sources

Integrating data from various sources allows for the creation of more cohesive and enriched datasets. Key advantages in this case include context enrichment, the addition of new connected data, and, as a result, improved model accuracy.

Context enrichment can occur through the addition of new information by incorporating new sources. For example, to describe a process or event, one might use news articles from a specific time period [19]. However, for a better understanding of the event, particularly in countries with strict media censorship, such data should be expanded using information from social networks, which tend to have lower levels of censorship.

The inclusion of new data sources, especially through the integration of fundamentally different knowledge bases, also contributes to a better understanding of the domain. For instance, biomedical research data can be enriched with clinical trial results [26], while statistical estimates can be complemented by studies from reputable sociology journals. The increase in volume and diversity of data leads to improved model accuracy, as evidenced by a number of studies, such as [26].

When combining data from various sources, as described above, we gain valuable new knowledge. However, this process can be complicated by differing data formats. A typical example is a situation where, on one hand, there is information describing cyberattacks presented in RDF format, and on the other hand, real data about vulnerabilities obtained through system logging of an organization's computer systems. In this case, the challenge lies not only in merging data of different natures but also in handling disparate formats. A complete implementation example of such a task is provided in Appendix B.

4.6 ALGORITHM FOR USING NATURAL LANGUAGE PROCESSING LIBRARIES AND SEMANTIC WEB TECHNOLOGIES FOR PATTERN IDENTIFICATION

The algorithm for enhancing named entity recognition (NER) accuracy using the Flair model can be described through the process of adaptive fine-tuning with knowledge extracted from RDF data, which has been previously obtained from open sources.

Initially, the Flair model, one of the most effective models for NER tasks (as demonstrated in our synthetic example), is used as the base model due to its architecture based on bidirectional language models. This model is already trained on large corpora and capable of recognizing named entities with high accuracy. However, to increase specificity and accuracy in a particular domain, the model requires fine-tuning with new, more specialized data.

Resource description framework (RDF) is a standard for describing metadata based on triples (subject-predicate-object), allowing for the structuring and semantic enrichment of information from various sources. Converting RDF data into a training set for the Flair model involves transforming triples into token sequences with annotations conforming to the CoNLL standard. This process requires semantic transformation, where entities from RDF data become named entities annotated in the text corpus, enabling their use for training. A graphical illustration of the algorithm is given in Figure 4.7.

The enhanced NER algorithm described above for analyzing the source text presented in Appendix A demonstrated better results compared to the initial data. In this analysis, three language models were used – BERT, RoBERTa, and Flair – each evaluated both with and without fine-tuning. The results of named entity extraction are summarized in Table 4.2.

4.7 DISCUSSION

4.7.1 Interpretation of Results

The results presented in Table 4.2 indicate that the Flair model, trained using domain-specific data extracted from RDF, demonstrated the best performance. Additionally, analyzing the results reveals a general trend of increasing NER accuracy for each model, ranging from 4% to 9%, when additional data is utilized. The highest efficiency of the fine-tuned Flair model, in our opinion, can be attributed to two main factors. First, Flair is trained on large corpora and already exhibits a high degree of accuracy, with fine-tuning not diminishing its initial precision but rather extending its capabilities. Second, using data in RDF format not only adds valuable new information but also links it to the original data due to the highly structured nature of this format. It is evident that the advantage of RDF as a data source lies in its high level of structuring, which also imposes limitations on its widespread practical application. For example, preparing RDF-formatted data for training requires significant time and effort, as these datasets can have complex hierarchies. Moreover, different tasks require different datasets for training. For instance, the fields of medicine and cybersecurity demand distinct sets of preprocessed and annotated information, which in turn leads to the challenge of collecting and properly processing primary data. This brings us to the issue of a model's ability to generalize, highlighting the significant performance drop a model may experience when applied to potentially new data from a different domain. All of the above emphasizes the ongoing challenge of improving the quality of automated data annotation to enable its transferability to new tasks.

4.7.2 Current Challenges and Limitations in Text Analysis Using Natural Language Processing and Semantic Web Technologies

At the current stage of NLP and semantic web development, issues related to ambiguity and contextuality of data are particularly relevant. Regarding word and phrase ambiguity, a key challenge arises in training NLP models, as an incorrectly identified

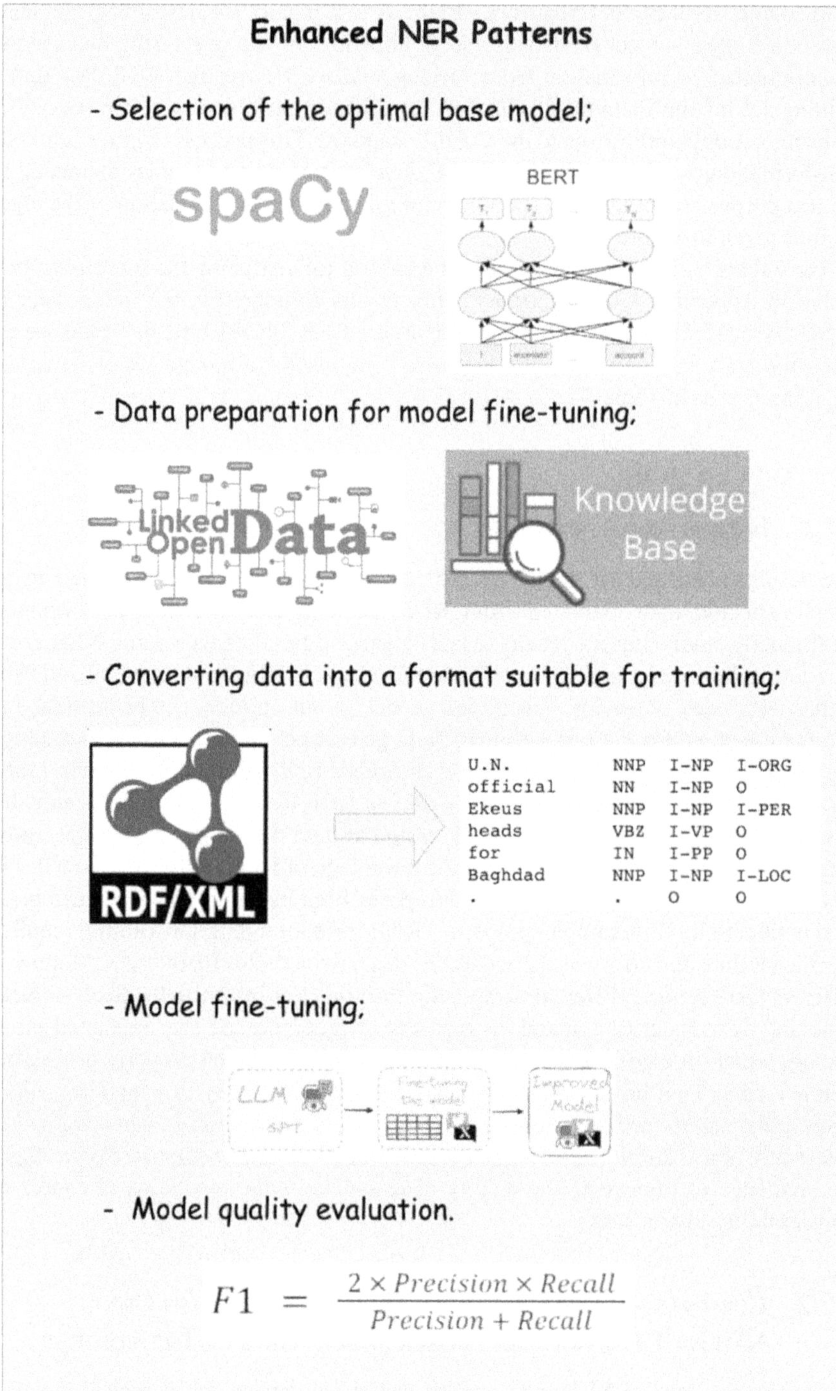

FIGURE 4.7 Enhanced Named Entity Recognition Algorithm.

TABLE 4.2
Named Entity Recognition Results

Type of NER Technology	F1 Score
BERT	0.65
RoBERTa	0.66
Flair	0.70
BERT with fine-tuning	0.69
RoBERTa with fine-tuning	0.73
Flair with fine-tuning	0.79

term reduces the overall effectiveness of the model. A potential solution to this problem is the use of word sense disambiguation (WSD) methods, which rely on contextual cues to determine the correct meaning of a word. However, their accuracy remains low due to the complexity of language and the inherent nature of ambiguity. Deep learning models, such as BERT and GPT, demonstrate significant improvements in resolving ambiguity due to their ability to consider context. Nevertheless, these models require large amounts of data and computational resources [27].

In terms of contextuality, challenges also arise in interpreting textual data, as determining the context of the current information can be problematic. This difficulty often stems from the inability to adequately account for cultural differences, existing world knowledge, and various other nuances. Contextual embedding technologies and memory-based models, such as LSTM and Transformer, address this issue. Contextual embeddings use neighboring words to determine the meaning of a word. However, despite their success, they are still limited in understanding complex and long-term contexts. Memory-based models – architectures like LSTM and Transformer – use attention mechanisms to retain context. However, they may lose important information over long distances, limiting their application for complex texts [28].

4.8 CONCLUSION

During the conducted work, a comprehensive analysis was carried out on the use of NLP technologies and the semantic web for pattern recognition and named entity recognition tasks. Key aspects of morphological, syntactic, and semantic analysis were studied, and modern approaches to integrating machine learning with semantic technologies were explored. The implementation of semantic web technologies, such as ontologies and RDF, significantly improved the results of entity recognition and classification, especially in specialized domains. One of the central issues in developing and fine-tuning machine learning models is the selection of training data. The effectiveness and accuracy of models directly depend on the quality of the data used, making data preparation a critical stage. However, the process of collecting and annotating data for fine-tuning, especially in specific domains, remains labor-intensive and prone to subjectivity. Many tasks require the creation of specialized datasets, adding further complexity to model development and adaptation.

The analysis and fine-tuning of the Flair model based on RDF-extracted data showed a significant improvement in results compared to baseline models such as BERT and RoBERTa. This fact confirms the initial hypothesis about the feasibility of integrating semantic web technologies to improve the accuracy of named entity recognition, as previously mentioned, once again highlighting the importance of the task of preliminary automated data annotation. Thus, the further development of the semantic web and ontologies will contribute to achieving more effective results in pattern identification and will facilitate the efficient resolution of specific applied tasks.

REFERENCES

1. Turing, A. (1950). Computing machinery and intelligence. *Mind*, *59*(236), 433–460.
2. Chen, D. (2024). Challenges of natural language processing from a linguistic perspective. *International Journal of Education and Humanities*, *13*(2).
3. Manning, C. D., Raghavan, P., & Schütze, H. (2008). *Introduction to information retrieval*. Cambridge University Press.
4. Porter, M. F. (1980). An algorithm for suffix stripping. *Program*, *14*(3), 130–137.
5. Toutanova, K., Klein, D., Manning, C. D., & Singer, Y. (2003). Feature-rich part-of-speech tagging with a cyclic dependency network. In *Proceedings of the 2003 conference of the North American chapter of the association for computational linguistics on human language technology – Volume 1* (pp. 173–180). Association for Computational Linguistics.
6. de Marneffe, M. C., MacCartney, B., & Manning, C. D. (2006). Generating typed dependency parses from phrase structure parses. In *Proceedings of LREC* (Vol. 6, pp. 449–454). European Language Resources Association.
7. Collins, M. (1997). Three generative, lexicalised models for statistical parsing. In *Proceedings of the 35th annual meeting of the association for computational linguistics* (pp. 16–23). Association for Computational Linguistics.
8. Nadeau, D., & Sekine, S. (2007). A survey of named entity recognition and classification. *Lingvisticae Investigationes*, *30*(1), 3–26.
9. Palmer, M., Gildea, D., & Kingsbury, P. (2010). Semantic role labeling. *Synthesis Lectures on Human Language Technologies*, *3*(1), 1–103.
10. Mitkov, R. (2002). *Anaphora resolution*. Routledge.
11. Heath, T., & Bizer, C. (2011). Linked data: Evolving the web into a global data space. *Synthesis Lectures on the Semantic Web: Theory and Technology*, *1*(1), 1–136.
12. Bechhofer, S., van Harmelen, F., Hendler, J., Horrocks, I., McGuinness, D. L., Patel-Schneider, P. F., & Stein, L. A. (2004). *OWL web ontology language reference*. World Wide Web Consortium (W3C). https://www.w3.org/TR/owl-ref/
13. Horrocks, I., Patel-Schneider, P. F., Bechhofer, S., & Tsarkov, D. (2004). OWL rules: A proposal and prototype implementation. *Web Semantics: Science, Services and Agents on the World Wide Web*, *3*(1), 23–40.
14. Ashburner, M., Ball, C. A., Blake, J. A., Botstein, D., Butler, H., Cherry, J. M., . . . & Sherlock, G. (2000). Gene ontology: Tool for the unification of biology. *Nature Genetics*, *25*(1), 25–29.
15. Gene Ontology Consortium. (2004). The gene ontology (GO) database and informatics resource. *Nucleic Acids Research*, *32*(suppl_1), D258–D261.
16. Aviad, A., Węcel, K., & Abramowicz, W. (2015). The semantic approach to cyber security: Towards ontology-based body of knowledge. In N. Abouzakhar (Ed.), *Proceedings of the 14th European conference on cyber warfare and security (ECCWS)* (pp. 328–336). Academic Conferences and Publishing International Ltd.
17. Harris, S., & Seaborne, A. (2013). *SPARQL 1.1 query language*. World Wide Web Consortium (W3C). https://www.w3.org/TR/sparql11-query/

18. Sikos, L. F. (2023). Cybersecurity knowledge graphs. *Knowledge and Information Systems*, *65*, 3511–3531. https://doi.org/10.1007/s10115-023-01860-3
19. Bizer, C., Heath, T., & Berners-Lee, T. (2009). Linked data – The story so far. *International Journal on Semantic Web and Information Systems (IJSWIS*, *5*(3), 1–22.
20. Šimko, J., & Bieliková, M. (2014). State-of-the-art: Semantics acquisition and crowdsourcing semantic acquisition. In P. Křemen, F. Kouba, & O. Škrabal (Eds.), *Semantic acquisition games* (pp. 21–41). Springer. https://doi.org/10.1007/978-3-319-06115-3_2
21. James, G., Witten, D., Hastie, T., & Tibshirani, R. (2013). *An introduction to statistical learning: With applications in R*. Springer.
22. Bishop, C. M. (2006). *Pattern recognition and machine learning*. Springer.
23. Aggarwal, C. C. (2015). *Data mining: The textbook*. Springer.
24. Goodfellow, I., Bengio, Y., & Courville, A. (2016). *Deep learning*. MIT Press.
25. Antoniou, G., & van Harmelen, F. (2004). *A semantic web primer*. MIT Press.
26. Belleau, F., Nolin, M. A., Tourigny, N., Rigault, P., & Morissette, J. (2008). Bio2RDF: Towards a mashup to build bioinformatics knowledge systems. *Journal of Biomedical Informatics*, *41*(5), 706–716.
27. Devlin, J., Chang, M. W., Lee, K., & Toutanova, K. (2019). BERT: Pre-training of deep bidirectional transformers for language understanding. In *Proceedings of the North American chapter of the association for computational linguistics: Human language technologies (NAACL-HLT)*. Association for Computational Linguistics.
28. Vaswani, A., Shazeer, N., Parmar, N., Uszkoreit, J., Jones, L., Gomez, A. N., . . . & Polosukhin, I. (2017). Attention is all you need. In *Advances in Neural information processing systems (NeurIPS)*. Curran Associates, Inc.

Appendix A: Textual Data for Analysis

In recent years, cybersecurity has become a critical concern for organizations and governments worldwide. In October 2021, a significant data breach occurred at Facebook, affecting over 500 million users. The breach was traced back to a group of hackers operating out of Berlin, Germany. This incident prompted the European Union to strengthen its cybersecurity policies, leading to the adoption of the Cybersecurity Act in 2022. The United States, through the Department of Homeland Security (DHS), also launched a comprehensive cybersecurity initiative in 2023 aimed at protecting critical infrastructure.

Companies like Microsoft and Apple have been at the forefront of developing advanced security solutions. In June 2023, Microsoft announced a partnership with the National Security Agency (NSA) to enhance cloud security protocols. Meanwhile, Apple's new iOS update included enhanced encryption features designed to prevent unauthorized access to user data. These measures came after a series of ransomware attacks targeting healthcare institutions across the United States, including a significant attack on the New York Presbyterian Hospital in July 2022.

The attacks were believed to be orchestrated by a Russian cybercriminal group known as DarkSide, which had previously targeted Colonial Pipeline in May 2021, causing widespread fuel shortages along the East Coast. The Federal Bureau of Investigation (FBI) and Interpol launched an international operation to track down and apprehend members of the group. By September 2022, key members of DarkSide were arrested in Moscow, Russia, in a joint operation with the Russian Federal Security Service (FSB).

In addition to ransomware, phishing remains one of the most common and effective methods used by cybercriminals. In March 2023, Google reported that it had blocked over 100 million phishing attempts targeting Gmail users. The report highlighted that phishing attacks had become increasingly sophisticated, with attackers impersonating organizations like PayPal and Bank of America to steal sensitive information.

Moreover, the rise of the Internet of Things (IoT) has introduced new vulnerabilities, with devices from manufacturers like Samsung and Huawei being targeted in various cyberattacks. In April 2023, the United Kingdom's National Cyber Security Centre (NCSC) issued a warning about potential threats to IoT devices, urging consumers to update their security settings regularly.

To combat these growing threats, governments and private companies have been investing heavily in cybersecurity research. The Massachusetts Institute of Technology (MIT) has launched a new research center focused on developing artificial intelligence (AI) solutions for detecting and preventing cyberattacks. This initiative, funded by a $100 million grant from the US Department of Defense (DoD), aims to create next-generation security tools capable of identifying threats in real-time.

Cybersecurity remains a dynamic field, with new challenges emerging as technology evolves. As organizations continue to digitize their operations, the need for robust cybersecurity measures becomes increasingly urgent. By staying ahead of threats and adopting proactive security strategies, we can protect sensitive data and maintain trust in the digital world.

This text includes numerous named entities, such as organizations (e.g., Facebook, Microsoft, Apple, NSA, and FBI), locations (e.g., Berlin, Germany, the United States, and Russia), and other specific terms related to cybersecurity. It should serve as a useful basis for testing named entity recognition models.

Appendix B: Example Code for Algorithm

DatainRDF

```
@prefix sec: <http://example.org/security#> .
@prefix sys: <http://example.org/system#> .
<http://example.org/attack1> a sec:CyberAttack ;
sec:attackName "Ransomware Attack" ;
sec:affectedSystem<http://example.org/system1> ;
sec:date "2023-07-15" ;
sec:severity "High".
<http://example.org/attack2> a sec:CyberAttack ;
sec:attackName "Phishing Attack" ;
sec:affectedSystem<http://example.org/system2> ;
sec:date "2023-08-02";
sec:severity "Medium".
<http://example.org/system1> a sys:InformationSystem ;
sys:systemName "Corporate Network".
<http://example.org/system2> a sys:InformationSystem ;
sys:systemName "Email Server".
```

Data in CSV

```
vuln_id,vuln_description,related_attack_id
CVE-2023-45678,Remote Code Execution, http://example.
org/attack1
CVE-2023-12345,Email Spoofing Vulnerability, http://ex-
ample.org/attack2
from rdflib import Graph
rdf_data = """
@prefix sec: <http://example.org/security#> .
@prefix sys: <http://example.org/system#> .
<http://example.org/attack1> a sec:CyberAttack ;
sec:attackName "Ransomware Attack" ;
sec:affectedSystem<http://example.org/system1> ;
sec:date "2023-07-15" ;
sec:severity "High".
<http://example.org/attack2> a sec:CyberAttack ;
sec:attackName "Phishing Attack" ;
sec:affectedSystem<http://example.org/system2> ;
sec:date "2023-08-02" ;
sec:severity "Medium".
```

```
<http://example.org/system1> a sys:InformationSystem ;
sys:systemName "Corporate Network".
<http://example.org/system2> a sys:InformationSystem ;
sys:systemName "Email Server".
"""
g = Graph()
g.parse(data=rdf_data, format="turtle")
import csv
from rdflib import URIRef, Literal, Namespace
vuln = Namespace("http://example.org/vulnerability#")
g.bind("vuln", vuln)
with open('vulnerabilities.csv', newline='') as csvfile:
reader = csv.DictReader(csvfile)
for row in reader:
vuln_uri = URIRef(f"http://example.org/{row['vuln_
id']}")
attack_uri = URIRef(row['related_attack_id'])
g.add((vuln_uri, RDF.type, vuln.Vulnerability))
g.add((vuln_uri, vuln.vulnDescription, Liter-
al(row['vuln_description'])))
g.add((vuln_uri, vuln.relatedAttack, attack_uri))
print(g.serialize(format="turtle").decode("utf-8"))
```

5 Futuristic Approaches in Conversational AI for the Cloud Platform

Panem Charanarur, Srinivasa Rao Gundu, Aditya Mohan Srivastava, H.K. Pratihari, Naveen Kumar Chaudhary, and Laishram Hemanta Singh

5.1 INTRODUCTION

Conversational AI refers to various techniques for training computers to have realistic conversations with people. Natural language processing (NLP) technology has led to a variety of machine learning models that can understand a wide range of inputs and participate in sophisticated debates. One popular form of conversational AI is chatbots, which use NLP to interpret user input and carry on discussions as illustrated in Figure 5.1. These AIs can be used in voice assistants, customer service chatbots, and virtual assistants.

Customers expect businesses to provide consistent high-quality customer service across various channels, and intelligent automation is essential to maintain these standards. AI can achieve better customer experience, satisfaction, loyalty, and lifetime value through "near-human" dialogues.

Open-domain conversational AI is used for free-flowing conversations on various subjects, such as leisure or task-oriented tasks. ELIZA, a conversational AI, was designed creatively and has been found to form personal ties with employees. SoTA conversational systems have lately been the focus of implementation efforts by a variety of open-domain conversational artificial intelligence systems. Through an analysis of the most recent open-domain SoTA conversational systems, we have identified a number of challenges that will be faced by "human" players in the "imitation game." Conversational artificial intelligence conversations that make use of low-resource languages are now being investigated from the standpoint of gender ethics [1].

5.2 CONVERSATIONAL FEATURES OF HUMANS

In social situations, one person may initiate a conversation, with the interviewer serving as the conversational leader. However, mixed or rotating initiative in conversational

 DOI: 10.1201/9781003512820-5

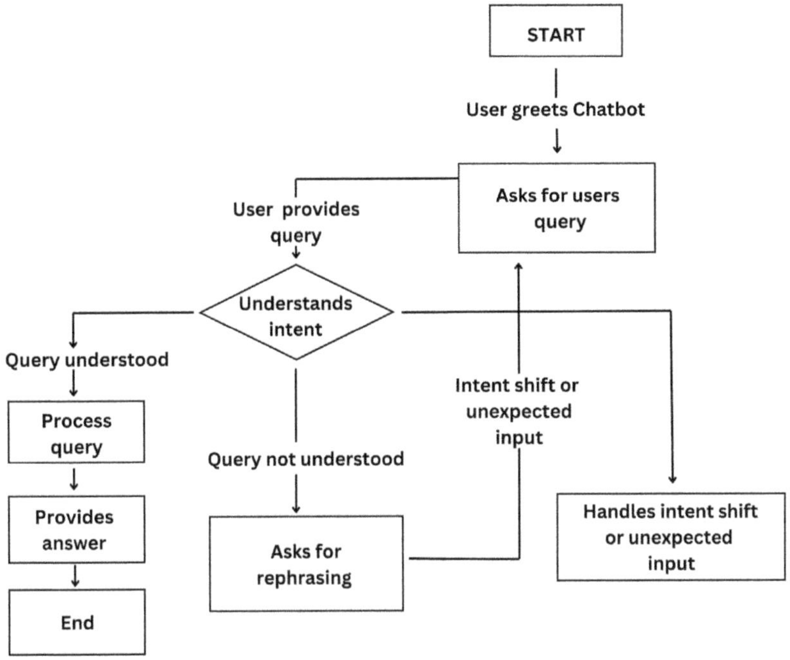

FIGURE 5.1 Flowchart of a Chatbot System.

systems is difficult. Human interactions are unique due to the number of speakers, the order of turns, the size of turns, the uncertainty of conversation duration, the variability in presenters and attendees, and the unpredictable frequency of each turn. A turn-based strategy is one option to address these challenges [2].

5.3 MORALITY IN ARTIFICIAL INTELLIGENCE-BASED CONVERSATION

Conversational AI raises ethical concerns due to its potential to be cruel or dehumanizing under certain conditions. However, deontological ethics prioritize objectivity and adhering to regulations like GDPR is crucial for responsible research. Conversational AI faces challenges in privacy, toxic language, and unintentional discrimination. Data privacy is vital as the security of the public may be jeopardized if important information is leaked. Debiasing methods and anonymization techniques are being used to address these issues. Finding a balance between chatbot usefulness and users' ethical concerns is challenging. Some question the inclusion of women's voices in technology, but it is widely accepted that more chatbots are aimed towards females than males. A study of 1,375 chatbots found the majority to be female [3].

5.4 ADVANTAGES OF ARTIFICIAL INTELLIGENCE-BASED CONVERSATION

The investigation of open-domain conversational artificial intelligence has garnered a great deal of interest for reasons that are not hard to understand. In this article, we will go over some of the advantages:

ELIZA and Kuki, who subsequently became known as Mitsuku, got along quite well despite their early doubts. It was reported that a significant number of employees had a greater sense of comfort after engaging in private talks with the chatbot.

Your assistance is much appreciated by visually impaired individuals as well as those with other disabilities. Conversational artificial intelligence, when combined with speech-to-text (STT) and text-to-speech (TTS) technologies, has the potential to be a lifeline for those who are unable to navigate their settings on their own. It is possible that our newly published IR approach may one day make it possible for conversational AI to get current information on specialized subjects [4].

Instructional materials that are straightforward, to the point, and efficient. It is essential to make observations that are plain and short as it is hard to forecast the nature of a debate or the length of anything that will be discussed.

The automation of high-quality data transmission between devices is a very important aspect, particularly for programming languages that have limited resources. It is possible for autonomous machine talks to create high-quality data, which might assist relieve the scarcity of data resources for languages with limited resources. These discussions can encompass a broad variety of topics and entities.

To replicate mental and psychological therapies for people, it is possible to utilize the findings of controlled trials as a basis. The purpose of these studies was to alter the correlations between inputs and outputs.

5.5 COMPONENTS OF ARTIFICIAL INTELLIGENCE-BASED CONVERSATION

Conversational AI consists of five core components: NLP helps computers understand spoken language and communicate with humans naturally, while Machine Learning improves through observation and experience. Text Analysis helps identify subjects, objects, and word categories in written documents, while computer vision analyzes visual input and emotions. Speech recognition software decodes spoken language and individual sounds, recognizing human voice and transcribing it into text. Overall, conversational AI aims to enhance human speech understanding and communication [5].

5.6 HOW CONVERSATIONAL ARTIFICIAL INTELLIGENCE ACTUALLY FUNCTIONS?

Automatic speech recognition (ASR) is a conversational AI system that translates spoken words into text using machine learning and deep neural networks. It uses natural language processing (NLP) to convert free-form text into a more formalized format for analysis. An AI model predicts the optimal answer based on user intent and training data. natural language understanding (NLU) analyzes data based on grammar, meaning, and context, while natural language generation (NLG) infers the best course of action for answering human questions.

5.7 ACHIEVING CONVERSATIONAL ARTIFICIAL INTELLIGENCE

The optimal approach to developing conversational AI will vary from company to company, depending on their own requirements and use cases; therefore, there's no universally correct response to this topic. Nonetheless, here are a few things to keep in mind while developing chatbots:

The first step in developing conversational AI is gaining an in-depth familiarity with your organization's unique use cases and requirements. In other words, what are you hoping to accomplish with your chatbot? How natural a conversationalist do you want it to be? Which metrics do you need to keep an eye on? The ideal strategy for developing your chatbot may be determined after you've defined these needs.

Second, choose the appropriate platform and toolkit: There are several options for developing conversational AI. Each platform has advantages and disadvantages; choose the one that works best for you.

Third, construct a prototype; when you've determined your needs and settled on a platform, you can begin working on it in earnest. Before releasing your chatbot to consumers, it is recommended that you create a prototype for testing purposes.

Once your prototype is complete, you may go on to step four: deploying and testing your chatbot. It's important to do a pilot test with a select set of people before releasing it to the general public.

As a last stage, optimize and enhance your chatbots on a consistent basis. Modifying the algorithms, incorporating user suggestions, and gathering data are all viable options [6].

5.8 INCORPORATION OF CONVERSATIONAL ARTIFICIAL INTELLIGENCE

Conversational artificial intelligence (CAI) can be implemented through natural language processing (NLP), converting human-written content into a computer-readable format. This technology can automate tasks like customer service responses. Explainable AI (XAI) developments make it easier for future robots to use AI. It is possible to optimize computers and enhance QoS via the strategic application of AI, ML, and DL in computer resource management. The development of new computer systems including cloud, fog, edge, serverless, and quantum computing has helped push the field of computer science and problem-solving forward. More young people are launching their own firms as a result, which is generating a surge of computing power that has the potential to speed up the expansion of scientific knowledge [7].

5.9 CONVERSATION DESIGN AND ARTIFICIAL INTELLIGENCE ENHANCING THE USER INTERACTION

Conversation design for chatbots and voice bots is a complex, human-led process that requires intent handling and determination. Machine learning models are trained using human interactions to improve their understanding of meaning. Human analysts and contact center agents use behavioral, corporate, and external data to forecast purpose and resolve issues more accurately. Conversational data is mined using unsupervised ML methods to identify trends in communication flow. Intent sequences are necessary for bot design to automate conversations, as bots may need to escalate to human agents if they cannot manage multiple meanings. Conversational AI is used in customer service, IT help desks, marketing, and sales [8].

5.10 CONVERSATIONAL ARTIFICIAL INTELLIGENCE AND CHATBOT DIFFERENCES

The subfield of artificial intelligence known as conversational AI is used in the construction of chatbots, which are computer programs that are able to carry on conversations using natural language. It is possible to give these bots instructions to carry out a variety of jobs in line with procedures that have been established or scripts that have been produced by people. Integration, escalation management, sophisticated analytics, intent management, language processing, response generation, and omnichannel user experience are some of the features that are included in this implementation. Chatbots provide you the ability to communicate with them via the use of text, speech, or a combination of the two. The ability to mimic human speech is only one of their many characteristics. They aim to develop a conversational self-service experience that is scalable, and they plan to do this by using a number of different digital and spoken channels. An option for a chatbot is to be powered by artificial intelligence (AI), or it might be built by humans [9].

5.11 DIFFICULTIES IN CONVERSATIONAL ARTIFICIAL INTELLIGENCE

Conversational AI has become a valuable tool for businesses, but it faces challenges such as developing natural language processing (NLP) skills, understanding human interactions, creating user-friendly experiences, integrating AI systems into existing businesses, protecting customer information, and adhering to laws and regulations. Building a robust conversational AI platform that supports local languages is crucial for global leaders. Developing AI-based conversational applications requires ongoing human-in-the-loop knowledge for managing conversations and designing dialogue. Finding suppliers with comprehensive solutions for AI-based applications is challenging. Explainable AI, which uses data science capabilities, helps explain the "why" behind model actions, boosting trust in the platform's accuracy. As customer demands evolve, updating automated discussions may be costly, and IT specialists, multi-persona models, or coding experts may be required [10].

5.12 ADVANCED CONVERSATIONAL ARTIFICIAL INTELLIGENCE

Through the utilization of advanced artificial intelligence, sentiment analysis, deep learning, natural language processing, and intent prediction, conversational AI is able to enhance customer interactions, decrease the amount of time required for resolution, and increase return on investment. During non-transactional interactions, sentiment analysis makes use of pre-built social and sentiment models to prioritize and escalate to agents as necessary. Natural language understanding (NLU) enables bots to respond with a human-like resemblance, while sentiment analysis is used to prioritize and escalate. Deep learning is a subfield of machine learning that attempts to simulate the way in which the human brain acquires knowledge and adjusts to novel circumstances. Voice synthesis that is powered by artificial intelligence is used to improve intent models through interactions between AI models, chatbots, and human agents. Intent prediction allows businesses to obtain a better understanding of the intentions of their customers and to give replies that are more pertinent to those intentions. A wide variety of domains, use cases, and vertical requirements are anticipated from conversational artificial intelligence systems [11].

5.13 UNDERSTANDING THE ROLE OF ARTIFICIAL INTELLIGENCE AND MACHINE LEARNING IN THE FUTURE OF COMPUTING

The Autonomic Computing Initiative (ACI) at IBM was an early effort to design computers that could accomplish predefined goals with little oversight from human operators. Cognitive science and the study of the nervous system, which is capable of carrying out several processes in parallel with little interference, were major influences on the work. Autonomous computer systems are designed to function independently, much like the human body. Self-adaptive systems, also known as self-configuration, are designed to perform tasks such as initialization, enhancement, defense, and maintenance without human intervention. Self-configuration maximizes efficiency by minimizing wasteful use of scarce resources, while self-improving systems successfully perform assigned tasks. Self-protection is achieved when a system can protect itself from outside interference. Self-healing systems diagnose malfunctions, determine their causes, and implement solutions. Self-managing and self-repairing systems should require zero human intervention beyond initial setup and regular upgrades [12].

Effective approaches to achieve these goals include the autonomic manager (AM), which gathers data from sensors and consults a small database of rules. AI and machine learning can be used to foster spontaneous routines, analyze workload patterns, and minimize model error. Control theory can also be used to improve autonomous self-management. Feedback-based control is a popular technique for giving computers autonomy. Self-organizing systems include cellular automata, genetic algorithms, and particle swarm optimization [13].

5.14 PURPOSE OF THE STUDY

Autonomic computing, incorporating AI/ML methods, is increasingly being used in various computer paradigms such as cloud, fog, edge, serverless, and quantum computing. This approach maximizes search efficiency by expanding parameter space and is particularly effective when used behind the scenes. P2P networks are already self-sufficient, and online services and data center management tools are incorporating self-management features. However, schedulers and process planners may not always have real-time input. As systems become more complex and interconnected, manual management and customization will become difficult and costly [14].

5.15 OBJECTIVES OF THIS STUDY

IBM outlines eight criteria for AI and ML-powered autonomous systems in computing: self-awareness, adaptability, prediction-making, self-correction, security, communication, and anticipating demands. These systems must be self-aware, resourceful, and capable of detecting and responding to threats. As the concept of "The Internet of Everything" (IoE) develops, interconnected networks will allow for communication between computers, making AI and machine learning essential for self-adaptive systems. Large corporations are investing heavily in AI and ML research to respond to changing demands in a dynamic environment [15].

5.16 WHY ARTIFICIAL INTELLIGENCE AND MACHINE LEARNING ARE CRUCIAL FOR FUTURE COMPUTING?

Autonomous computing offers numerous benefits, including reduced total cost of ownership, reduced maintenance costs, and increased reliability in IT infrastructure. It can adopt company strategy and adapt to environmental changes, leading to higher-order benefits. AI-based autonomous computing streamlines system administration, enhances computer performance, and allows for distributed workloads. Cloud-based data storage and processing can save money and increase output by automating routine tasks and data analysis without human intervention. Integrating AI technology with Google Cloud Stream analytics allows for real-time personalization, anomaly detection, and management scenario prediction. AI-based data protection measures are crucial for cloud-based app deployments, ensuring the security of sensitive data.

5.17 A PROSPECTIVE MODEL FOR FUTURE COMPUTING SYSTEMS

Gateway devices are crucial for connecting IoT/edge devices and end users to computer systems, ensuring the functionality of AI and autonomous services. Distributed systems like IoT platforms need to manage resources and schedule operations in a flexible and error-tolerant manner. IBM's autonomous computing framework emphasizes self-healing, self-configuration, self-protection, and self-optimization capabilities. AI-based systems can use data sources to construct issue models, allowing proactive problem analysis and maintenance. Self-protection aims to protect the system from potential attacks. Prediction systems based on AI can recognize communication vulnerabilities. AI and ML-based adaptive scheduling is useful for applications dealing with large amounts of data. The knowledge base contains significant data, and the management layer for services includes database managers, security managers, serverless managers, application managers, and placement modules [16].

5.18 MODEL WITHIN ITS OWN AUTONOMOUS CONTROLS

IBM's next-generation architecture is based on its autonomic computing paradigm, which emphasizes self-optimization, self-protection, self-healing, and re-configuration. This approach involves monitoring, analysis, planning, and execution, with AI and ML models playing a crucial role in addressing state synchronization issues in unstable

TABLE 5.1
Chatbot Evolution: From Basics to Intelligence

Traditional chatbots	AI-powered chatbots	
1. It has a rather simple structure.	1. Transactionally oriented	1. Intricate and situational
2. Basic question-and-answer devices	2. Capable of handling intricate conversations	2. It extends beyond just talk.
3. Be easily integrated	3. Compatible with a wide variety of back-end and legacy systems	3. Intelligent and aware of its surroundings
4. Be of restricted scope	4. Strategically oriented	4. Capable of learning and improvement over time
5. Require case-by-case instruction	5. Develop expertise in using several systems to accomplish a goal	5. Preemptively meeting users' demands
6. Need little in the way of follow-up work	6. Demand a lot of work towards the end	6. Involve a lot of work in the background

environments. Sensors provide quality-of-service measurements, while security agents help identify and mitigate threats. Self-healing and self-protection are achieved through hardening agents, minimizing attack surfaces in software, networks, and hardware.

The service-oriented architecture (SOA) stores and manages various types of data, including sensor data, system specifications, and AI/ML approaches. The management layer of SOA comprises a database manager, security manager, application manager, and placement module, which use AI technology to anticipate potential threats and protect employees.

Quantum computing is used in place of traditional computing to complete proof-of-work (PoW) and nonce calculations faster.

The integration of AI and cloud computing is a significant development in next-generation computing, improving operational efficiency, strategic focus, and decision-making. Hosting AI on the cloud offers advantages such as improved data management, automation, cognitive automation, cost savings, and the development of cutting-edge AI tools and accelerators [17].

5.19 NEXT-GENERATION COMPUTING WITH EXPLAINABLE AI (XAI)

Computer systems require sound judgment to ensure their reliability and ability to perform assigned tasks. Errors in judgment can have catastrophic consequences on finances and physical assets. Artificial intelligence and machine learning have improved resource allocation and energy generation, but their predictions are not yet relevant, explainable, or executable. Recent research focuses on achieving quality of service, and understanding explainable AI (XAI) and XAI tools and procedures is essential for making resource management choices in compliance with best practices. Proper implementation of XAI prediction models, such as forecasting resource and energy usage and deviations from service level agreements, is crucial for computers to become more useful, explicable, and executable. Leveraging techniques from deep learning can also help address these concerns. Overall, understanding explainable AI and its tools is crucial for achieving computer goals and ensuring the future of the computer industry [18].

5.20 POSSIBLE DANGERS INHERENT IN ARTIFICIAL INTELLIGENCE-BASED INTEGRATED COMPUTING

AI's success depends on extensive training, Internet connection, user data privacy, and trust management. Slow Internet connections can hinder machine learning and autonomous computing. Encrypting user data is crucial for fully autonomous AI systems. AI's complexity makes errors easy, necessitating trust and management. Data engineering techniques, including choosing representative datasets and conducting phases, are essential for defining AI model quality. The MAPE-K loop must handle these steps to ensure AI's success in computer systems [18].

5.21 CONCUSSIONS AND FUTURE PERSPECTIVES

The availability of vast amounts of data and computational resources is crucial for the implementation of AI, ML, and DL methods. Computers simplify this process and allow for the simultaneous training of complex models. Traditional data center resource management might see a resurgence in interest as a result of AI, ML, and deep learning applications. Improving resource management and maximizing service quality could be possible outcomes of developing explainable AI, which might lead to the widespread usage of AI in state-of-the-art computers.

5.22 STATEMENT OF FUNDING

This chapter is not funded by any organization.

5.23 CONFLICT OF INTEREST STATEMENT

We do not have any conflict of interest.

REFERENCES

[1] J. Zhu and J. Van Brummelen, "Teaching Students about Conversational AI Using Convo, a Conversational Programming Agent," *2021 IEEE Symposium on Visual Languages and Human-Centric Computing (VL/HCC)*, 2021, pp. 1–5, https://doi.org/10.1109/VL/HCC51201.2021.9576290.

[2] C. Deka, S. Sah, A. Shrivastava, M. Phukon and L. Routray, "Assessing a Voice-Based Conversational AI prototype for Banking Application," *2021 8th NAFOSTED Conference on Information and Computer Science (NICS)*, 2021, pp. 211–216, https://doi.org/10.1109/NICS54270.2021.9701536.

[3] P. Kulkarni, A. Mahabaleshwarkar, M. Kulkarni, N. Sirsikar and K. Gadgil, "Conversational AI: An Overview of Methodologies, Applications & Future Scope," *2019 5th International Conference On Computing, Communication, Control And Automation (ICCUBEA)*, 2019, pp. 1–7, https://doi.org/10.1109/ICCUBEA47591.2019.9129347.

[4] A. Fuad and M. Al-Yahya, "Recent Developments in Arabic Conversational AI: A Literature Review," *IEEE Access*, vol. 10, pp. 23842–23859, 2022, https://doi.org/10.1109/ACCESS.2022.3155521.

[5] A. Rajagopal and N. Vedamanickam, "New Approach to Human AI Interaction to Address Digital Divide& AI divide: Creating an Interactive AI Platform to Connect Teachers & Students," *2019 IEEE International Conference on Electrical, Computer*

and Communication Technologies (ICECCT), 2019, pp. 1–6, https://doi.org/10.1109/ICECCT.2019.8869174.
[6] X. Song and T. Xiong, "A Survey of Published Literature on Conversational Artificial Intelligence," *2021 7th International Conference on Information Management (ICIM)*, 2021, pp. 113–117, https://doi.org/10.1109/ICIM52229.2021.9417135.
[7] Y. Handrianto, R. Huang and Y. Shang, "Short Paper: TigerAware Assistant: A New Serverless Implementation of Conversational Agents for Customizable Surveys on Smart Devices," *2019 First International Conference on Transdisciplinary AI (TransAI)*, 2019, pp. 88–91, https://doi.org/10.1109/TransAI46475.2019.00023.
[8] M. Yousef and M. A. Torad, "A Treatise On Conversational AI Agents: Learning from Humans' Behaviour as a Design Outlook," *2019 International Conference on Electrical and Computing Technologies and Applications (ICECTA)*, 2019, pp. 1–4, https://doi.org/10.1109/ICECTA48151.2019.8959585.
[9] R. Tejwani, F. Moreno, S. Jeong, H. Won Park and C. Breazeal, "Migratable AI: Effect of Identity and Information Migration on Users' Perception of Conversational AI Agents," *2020 29th IEEE International Conference on Robot and Human Interactive Communication (RO-MAN)*, 2020, pp. 877–884, https://doi.org/10.1109/RO-MAN47096.2020.9223436.
[10] J. Chu, "Recipe Bot: The Application of Conversational AI in Home Cooking Assistant," *2021 2nd International Conference on Big Data & Artificial Intelligence & Software Engineering (ICBASE)*, 2021, pp. 696–700, https://doi.org/10.1109/ICBASE53849.2021.00136.
[11] C. Yildirim, "An Immersive Model of User Trust in Conversational Agents in Virtual Reality," *2021 Third International Conference on Transdisciplinary AI (TransAI)*, 2021, pp. 17–18, https://doi.org/10.1109/TransAI51903.2021.00011.
[12] J. Wall, "Keynote Speech 4: A Conversational AI Approach to Detecting Deception and Tackling Insurance Fraud," *2021 Tenth International Conference on Intelligent Computing and Information Systems (ICICIS)*, 2021, pp. 31–31, https://doi.org/10.1109/ICICIS52592.2021.9694118.
[13] S. DiPaola and Ö. N. Yalçin, "A Multi-layer Artificial Intelligence and Sensing Based Affective Conversational Embodied Agent," *2019 8th International Conference on Affective Computing and Intelligent Interaction Workshops and Demos (ACIIW)*, 2019, pp. 91–92, https://doi.org/10.1109/ACIIW.2019.8925291.
[14] L. Nie, F. Jiao, W. Wang, Y. Wang and Q. Tian, "Conversational Image Search," *IEEE Transactions on Image Processing*, vol. 30, pp. 7732–7743, 2021, https://doi.org/10.1109/TIP.2021.3108724.
[15] K. Xu, H. Wu, L. Song, H. Zhang, L. Song and D. Yu, "Conversational Semantic Role Labeling," *IEEE/ACM Transactions on Audio, Speech, and Language Processing*, vol. 29, pp. 2465–2475, 2021, https://doi.org/10.1109/TASLP.2021.3074014.
[16] A. Mehta, Y. Kunjadiya, A. Kulkarni and M. Nagar, "Exploring the Viability of Conversational AI for Non-Playable Characters: A Comprehensive Survey," *2021 4th International Conference on Recent Trends in Computer Science and Technology (ICRTCST)*, 2022, pp. 96–102, https://doi.org/10.1109/ICRTCST54752.2022.9782047.
[17] J. Tian, Z. Tu, Z. Wang, X. Xu and M. Liu, "User Intention Recognition and Requirement Elicitation Method for Conversational AI Services," *2020 IEEE International Conference on Web Services (ICWS)*, 2020, pp. 273–280, https://doi.org/10.1109/ICWS49710.2020.00042.
[18] B. Basnyat, N. Singh, N. Roy and A. Gangopadhyay, "Vision Powered Conversational AI for Easy Human Dialogue Systems," *2020 IEEE 17th International Conference on Mobile Ad Hoc and Sensor Systems (MASS)*, 2020, pp. 684–692, https://doi.org/10.1109/MASS50613.2020.00088.

6 Harnessing the Power of Artificial Intelligence to Enhance Digital Forensics

Raunak Sood

6.1 INTRODUCTION

To gain a full understanding of how AI is revolutionizing the field of digital forensics, it is quite important that we firstly understand what artificial intelligence (AI) is. AI is a branch of computer science whose major objective is the creation of machines that can solve problems like a human being by applying the same intelligence that a human being might possess. AI is not a new concept at all; it has penetrated our daily lives, and its usage has increased dramatically over the past few years. For example, when we access websites like Netflix or Amazon, an AI-based chatbot pops up to assist the user, and more advanced and intelligent virtual assistants like Siri and Alexa seem to be present across every corner of the globe.

Digital forensics (DF) is majorly concerned with finding, conducting analysis, and understanding the electronic record on devices and networks. The aim and objective of DF is to carry out a methodical investigation by helping forensic scientists conduct investigations into electronic records and devices, enabling them to handle more complex and difficult situations. Nowadays, cybercrimes are getting more and more complicated and sophisticated, making it extremely difficult to combat them in the borderless cyberspace. AI enables computers to analyze data from past cyberattacks, automatically learning and improving from that experience. Machine learning (ML) and its connected algorithms help AI conduct an analysis of patterns in the data that might be missed by a human being or may be time-consuming for a human to analyze and catch upon.

Currently, AI is stepping into the arena of network forensics, image and video forensics, and cryptocurrency forensics. The benefits associated with the usage of AI in DF is that AI is actively engaging in pattern recognition, anomaly detection, natural language processing, predictive analysis, and automation. Despite AI's inherent capabilities, its integration into digital forensics is not yet fully realized due to the extensive data demands for proper model training. Thus, there are many privacy concerns, biases in AI, and other potential concerns that could lead to results that are potentially not acceptable and misleading. Despite the challenges,

DOI: 10.1201/9781003512820-6

the future of AI is quite great if researchers are able to address the challenges and limitations of AI because, as AI is growing, it has become quite a significant tool for countering cybercrime.

6.2 USAGE OF ARTIFICIAL INTELLIGENCE IN ACQUISITION OF EVIDENCE

AI is a powerful tool when it comes to increasing the effectiveness and efficiency of digital forensic investigations. AI can help in the acquisition of evidence by improving the methods and techniques, which can be used for the analysis of evidence and the systematic prediction of cyberattacks by automating tasks incidental to collection, analysis, and reporting, identifying different patterns, and giving better insights. AI is useful in the acquisition of evidence because integrating AI and digital forensics can help in the following ways:

I. **Automating data collection:** Tracking down the source where the evidence has been generated and sometimes tracking down who the attacker is if AI steps and automates the data collection process. As AI can preserve and identify volatile data such as RAM data, which may be lost if the device is shut down or restarted, to quite an extent, AI can remind the investigating officer of such contingencies and then help forensics by helping them with methods for acquiring volatile data. [2]

II. **Prevention and Analysis:** AI will spend less time acquiring evidence compared to humans, but there is a caveat here that AI might not have access to the supporting files on the system; thus, it is likely to make errors. Therefore, it should be noted that files and evidence to investigate an attack is needed through human intervention. The extraction and analysis of data from encrypted devices is one example where encryption keys may not be known to the forensic analyst, but AI, going through its huge database, can generate encryption keys. [3]

III. **Malware Detection and Network Traffic Analysis:** For malware detection and network traffic analysis, the effectiveness of evidence detection and acquisition relies on AI algorithms. A key advantage of integrating AI with forensic tools is its capacity to analyze memory blocks that traditional tools cannot, thereby preventing intruders from hiding evidence. [4]

It is argued that though AI combined with DF is a powerful tool for the successful acquisition of evidence, it cannot completely replace human investigators. Therefore, to automate and conduct a more comprehensive analysis of crime scene evidence, it is crucial that we give more independence to AI. Where AI is given the freedom to generate and crunch on data which is publicly available, AI can prove to be helpful towards DF for the analysis of evidence and make it relatively easier for the forensic experts in conducting the analysis of evidence which has been acquired from the crime scene. [5]

Let's try to develop a working model for AI to help in the collection of evidence. Firstly, the AI will have to conduct a preliminary assessment of the file by checking the read-only status, clone, timestamp, and verification of the file and its metadata. Secondly, the AI should then process and detect network sniffing, process monitoring, honey trapping, intrusion detection, etc. If this hurdle is clear, then lastly, the AI should further proceed towards classifying all the collected data into three categories: structured, unstructured, and semi-structured data. This is a primary and most basic structure constructed here; if most of the basic functions are taken care of, then it will be much easier for the investigator because when conducting a forensic investigation, the most crucial thing is for the investigator is to avoid wasting time by clearly sifting through vast amounts of data. Here, AI steps in to clearly segregate the data, thus saving time for the investigator. It is argued that when AI is integrated with forensic tools, then it is possible that the efficacy of the tool can be improved because AI can give specific instructions that are objectively and specifically defined for achieving accurate and efficient goals, which completely values the standard intelligence systems present within the forensic tools. When the forensic tool receives instructions, AI can help the investigator with a preliminary analysis of the evidence collected. Of course, AI cannot replace human intellect in reaching a conclusion, but if AI helps in reaching a preliminary conclusion, then the investigator will be relieved to have a huge chunk of work done by AI. [6]

Another use for AI is in volatile memory evidence retrieval, but for doing so there is an extensive amount of work needed to train the AI system for collection of enough data. Now one of the possible approaches is that the AI can find similar patterns within the database it has been trained on through the machine learning process and then collect data on the crime scene, and the collected data will help in the analysis of the crime scene data. [7]

AI can be used to study the data collected from storage devices for analyzing and detecting information that has been collected from the devices, but here AI can also detect alterations in the data. When the forensic expert analyzes the data, AI can help in classifying the application of machine learning towards forensic tools by analyzing common characteristics in the data and handling problem classification for making and maintaining the cluster sizes, which, upon analysis, may give extremely accurate results. AI can analyze metadata, and if AI is able to make groupings of the meta data to get a perfect mapping, it may show how the data is acquired end it into various groupings, which might render it possible because of the generative capabilities of AI. [9]

Potential techniques that can integrate with AI:

1. Pattern recognition
2. Expert systems
3. Knowledge recognition

Now that AI has been integrated with the forensic too here comes in "smart acquisition" of data. For example, when a threatening email is sent to a VIP, the VIP will file a report with the police only after receiving the threatening email. If we apply a non-AI forensic tool to get an image from the storage device then it is crucial that the tool will image all the data, and the investigator will then have to extract all the files from that whole image. Let's assume a 1TB hard disk is going to be analyzed;

creating a forensic image is extremely time-consuming, and that very image has to be analyzed by some tool. Here, AI comes into the game where the AI will be trained on similar cases, AI will know from past experiences which data is relevant and of interest to the forensic investigator. As per the present case, AI will only look through emails, so the investigator will not have to concern himself with the multimedia files or even the software that has been installed on the machine. So, a "smart image" will be generated by AI. [11]

Now in the above example, how **smart acquisition work** is divided into three phases:

1. **Learning phase**: In this phase, the AI will first segregate the files and locations that have been used by the investigator for solving the case.
2. **Outside of Learning Phase:** Once the learning is over, the AI forensic tool will focus on the real-time situation at hand and look at the case. Now the AI forensic tool will engage and focus on those locations and files that are suitable to the requirements of the investigator.
3. **Checking and testing for integrity:** The integrity of digital evidence is extremely crucial when handling evidence, so it is crucial for AI to be trained in a way that only relevant data is acquired, so it is proposed that to overcome the integrity issue, it is crucial that approximately two copies of evidence are created, where the first copy is the forensic image of the disk and the second copy consists only of the relevant files that have been filtered by AI.

The challenge to this method is that it is going to be extremely time-consuming, but at the current level of technology, it is crucial that a well-technologically developed AI tool can help reduce the time in the next step, which is the analysis of evidence [12].

6.2.1 Case Analysis

VolMemLyzer [13]: This is a very famous forensic tool named Volatility Memory Analyzer. This tool is a part of memory forensics, which inspects malicious activities when the system is infected with malware. Memory analysis not only looks for and captures footprints but also collects important hidden code from the plain sight of the investigator. In simple words, this tool enables the analysis of the random-access memory or other types of volatile memory. This is a memory analysis tool for identifying specific areas of malware infection, so digital forensic analysts have been guided as to where they should focus their forensic examinations. One of the limitations of this tool is that it can memorize training samples that are given to it, but even after training, it is unable to produce correct results. So now, applying the concept of "smart acquisition" to improve the tool.

Step 1: **Learning Phase:** Here we will supply the AI-integrated version of VolMemLyzer with training modules that employ multiple classifiers so that a clear classification of the malware families is generated. We will use k-nearest neighbor and Random Forest as the main classifiers, and previous research suggests that these classifiers help better in classifying the malware functions.

Step 2: **Outside of the learning phase:** In this phase, since multiple classifiers have been used in the learning phase, we will now apply them to a normal real-time situation. Now, in this phase, the tool will analyze the functions, and the AI system, through its training, will overcome the inaccuracy in classification. The AI, having a larger database, will also look into the issue of fewer memory dumps after the execution of the malware, thus effectively overcoming the limitations of the tool itself. This is going to prove groundbreaking in the area of memory forensics.

Step 3: **Checking and testing for integrity:** Same as above.

Now we can see that the efficiency of the tool related to memory forensics will have increased quite a lot.

Let's see what the challenges of this testing hypothesis. The major challenge in memory forensics is that there is a huge amount of dependency on training modules. If there is a case that is unique and one of its kind, then this proposed framework might not work at all. It is difficult for a forensic investigator to train the AI-based tool with so much data, but this is possible at the manufacturer level. Thus, it is imperative that the tool must be used for memory forensics.

6.3 SAFEGUARDING PRIVACY IN ARTIFICIAL INTELLIGENCE-BASED DIGITAL FORENSICS

Digital forensics and artificial intelligence go hand in hand. There needs to be an interim stage where digital forensics meets artificial intelligence, but when this fated meeting takes place, there arises a demand for ensuring privacy-friendly forensics with the help of artificial intelligence. Therefore, there is a need for knowing and evolving the principles of usage of artificial intelligence in a privacy-friendly environment. [14]

But before engaging in the debate of ensuring the balance between privacy, digital forensics, and artificial intelligence, let us firstly delve into the recent developments of laws and attempted regulations surrounding artificial intelligence. A comparative study will be undertaken to know about the artificial intelligence regulations followed in various countries.

6.3.1 The United States

On 30 October 2023, US President Joe Biden took the first step in regulating AI by signing an executive order, thereby establishing new standards for the usage of AI. [15] The contents of the order are summarized as follows:

1. **Government Control**: The government of the United States poises to have control over all the AI being generated in the country. In compliance with the Defense Production Act, all companies that are trying to create and train AI models which might pose a threat to national security, economic stability, public health, and safety are duty-bound to give disclosure to the US Federal Government during the training process. The executive order further has made it compulsory for the declaration of safety assessment results conducted by independent teams, thus making AI more safer, more secure, and more reliable for consumers before it is launched on public platforms.

2. **Establishment of Regulatory Control:** To guarantee the safety, security, and reliability of AI, the executive order gives responsibility to various departments like the National Institute of Standards and Technology (for defining the standards for AI so that the safety of the AI can be audited before public release) and the Department of Homeland Security (addressing the threats and risks that AI might pose to the public).
3. **Safeguarding the public:** To safeguard the public, the order envisages that a cybersecurity initiative for the development of AI tools for identifying and assessing vulnerabilities in crucial software. The order directs the creation of a National Security Memorandum, which gives an outline about AI and the security associated with it. The chief of staff has the responsibility to draft and develop this memo for usage of AI in a safe, ethical, and secure manner.

   Now for the protection of the privacy of consumers in the United States, the executive order recognizes that AI poses a grave threat to the privacy of the general populace because companies train AI by exploiting the personal data of individuals. Therefore, the President, in the executive order, has called upon the US Congress to pass legislation protecting the privacy of Americans. The following principles have been recognized:

1. Development of privacy-preserving techniques for leveraging cutting-edge AI: The order proposes that the US government should spend more on privacy-friendly techniques like cryptography, including leveraging AI while safeguarding the privacy of the underlying data and funding the Research Coordination Network for getting breakthroughs in technological developments. It is the duty of the National Science Foundation for collaborating on these parameters.
2. The federal government will have to take up the mettle and strengthen guidelines for its own agencies to address the risks and information associated with AI, and attention will be given to commercial information, i.e., the personal data of individuals.
3. Formulation of guidelines for federal enforcement offices or agencies to assess the measures of techniques that are more privacy-friendly for deployment in AI systems; hence, this will lead to better safeguarding of the private data of Americans.

6.3.2 The European Union

On 28 September 2022, the EU came up with a draft directive (Directive) that affixed liability for non-contractual and AI fault-based civil claims, along with various rules for preserving and disclosing evidence where there is involvement of highly risky AI. The directive was further responsible for the establishment of the burden of proof and the presumptions arising during the usage of AI. [16] The salient features of the AI Directive are as follows [17]:

1. Courts are given the power to pass orders for giving information about their devices or systems to those persons or service providers who are seeking information for legal redressal against that service provider or user of AI.

The condition precedent is that the person who is requesting such information from the court has to firstly show that why does they think that the high-risk AI system is going to cause the alleged damage.

2. The principle of proportionality and necessity is adhered to here because the court has the duty to see that only proportionate information is being given to the requestor for supporting claims for damages. The legitimate interests of all parties are to be considered, including IPR protection, trade secrets, etc. If any of the documents are seen by the court to be a trade secret that might harm legitimate interests, the court has the utmost duty to maintain confidentiality.
3. The draft lays down circumstances in which the court has to presume a causal link between the fault of the service provider of the AI system and the user of that AI system, and then finally, there is a need to also see the output produced by the AI system and the consequent failure of the AI system to produce such outputs.
4. The draft legislation also mentions the consequences for non-compliance with the standards under the draft AI Directive and also talks about the presumption that might not apply if the court is going to find out that there is lack of sufficient evidence and a lack of a causal link between the same.

6.3.3 India

In India, there is no law at present that regulates AI, but it is the NITI Aayog that came out with a paper titled "National Strategy for Artificial Intelligence #AIforall." [18] An analysis of the recommendations of that paper is being conducted as follows:

1. The paper highlights privacy concerns with regard to AI. It suggests that the development of AI should take place in a marketplace where there is an audit-trail mechanism for addressing security and privacy issues. The paper does not agree with the suggestion of having a central nodal agency for facilitating access to companies for training AI on user data because users may not agree to the storage of their precious data someplace else. So, the paper advocates for a decentralized data marketplace, somewhat similar to blockchain technology, but with conditions for accessing that blockchain-like market such as traceability, price discovery mechanisms, regulatory compliance, and access control. This would enable AI development companies to train AI, but, in the marketplace, there will be controls over privacy and anonymization of persons. The proposed mechanism for realizing a decentralized marketplace based on decentralized ledger technology will allow for the development of privacy preservation.
2. **Principles on which AI should be based:**
    1. Fairness
    2. Transparency
3. **Suggestions on privacy issues include:**
    I. Establishing sectoral regulatory frameworks for AI and privacy: the paper has given examples of Japan and Germany where sector-specific legislations have been enacted for the regulation of AI and self-driving cars.

II. Establishing compliance with international standards, like the global initiative in ethics of autonomous and intelligent systems of the IEEE, which has a chapter that addresses '**personal data and individual access control in ethically aligned design.**'
III. Usage of tools for the assessment of data impact, which should be used by AI developers and encourage self-regulation by the companies themselves.

6.3.4 Comparative Analysis

On an analysis of the United States, EU, and Indian approaches, the points of similarity are listed below:

1. The objective is to preserve the privacy of the user.
2. AI should not train on user data without the consent of the user
3. Companies that are training AI should comply with sectoral regulations.
4. Privacy-preserving techniques should be evolved, and new mathematical models should be evolved to ensure that both users and companies at the manufacturing level coordinate with each other for training AI.
5. Transparency and fairness are needed so that no harm is caused to the user.

6.4 DIGITAL FORENSICS AND PRIVACY ISSUES

Digital forensics has major flaws with respect to the privacy of the user of the device in the case where the device is being investigated. Let us look at the technological challenges related to digital forensics and surrounding privacy issues, which are as follows [19]:

1. **Third-Party Privacy Exposure**: When the investigator accesses the device, he might get to know information about third parties who are not related to the accused under investigation, so the privacy issue of "express consent" of third parties comes into play.
2. **Cloud Computing:** In cloud computing, the jurisdiction and its privacy laws differ because the cloud can be accessed from anywhere and by any user and there is an international fragmentation of privacy laws; therefore, it is very difficult to come up with a uniform privacy framework across the world.
3. **Database Forensics:** The privacy issue in database forensics is the right of the user to protect access to that data, which is not relevant to the investigator. After the investigator accesses data, there should be secure erasure and that too within a specific time limit.
4. **Data Collection**: The investigator, when making a forensic copy, almost collects all the data, and there is no collaboration or cryptographic technique applied to keep user data private.
5. **Centralization of Data**: The investigator, when investigating a device, might exceed the mandate and scope of authorization of the investigation and further check out the interesting content on the device that is not at all relevant to the investigation.

TABLE 6.1
Points of Differentiation

Type	India	United States	European Union
Government control	States that there is no need for government control, but the government has a limited role where regulations for the usage of AI should be sector-specific.	The approach is more government-centric, where there is a need to share data with government authorities and companies with the standards set by the government.	The approach is that there is minimum government interference, and all the control for AI lies with the user and service provider of the AI company.
User	The user is the beneficiary, but instead, the user voluntarily enters a decentralized marketplace to share his data.	The user and his interests are majorly protected by the government.	The court or the judicial officer acts as the protector of the interests of the user.
Legal test	The DPDP Act governs the privacy of the user, and it is the consent of the user that matters the most.	No federal privacy law is in force; hence, for the time being, the government will make sure that AI trainer companies share all the data with the government, and the government will approve the AI-based products on a case-by-case basis.	The user can request the information which is being held by the AI trainer company, provided that significant damage has been caused to the user and the information will be granted by the court on grounds of necessity and proportionality.
Company developing AI product	The company just has to comply with sectoral regulations and privacy norms.	The company has to take federal government approval and lots of compliance.	The company must comply with privacy law and obtain user consent.
Nodal agency	No, because the marketplace for the development of AI is decentralized.	Yes, there are different agencies for the regulation of AI.	No, it is totally user-based.

6. **Not Judging the Value of Data Correctly**: Investigators might not know the value of the data they are investigating and might suspect the accused based on data that is not relevant. For example, if there is a case of electronic voyeurism via email, the investigator is to investigate only the stalking angle but might open the email containing voyeuristic material, including other attachments from the same person who sent voyeuristic messages.
7. **Criminal Use of Digital Forensics**: Internet of Things (IoT) and their systems, like the systems used for home monitoring, are at risk of misuse by criminals using even a little bit of digital forensics to unlawfully know the location of individuals.

8. **Finally, there is a lack of tools for privacy management in digital forensics**: Major tools like SleuthKit, FTK, and EnCase currently provide very little support for tracking privacy issues present during forensic investigations.

Now, a data crunching algorithm further threatens user privacy in digital forensics; hence, the authors have come up with solutions to the problems and issues raised. [20]

6.5 ARTIFICIAL INTELLIGENCE IN DIGITAL FORENSICS

AI is a very new field that is having a major impact on digital forensics. The majority of issues which are evolving in the overlap between AI and digital forensics is that there is a large amount of data that is being stored in electronic format, and there are many technical problems to take care of when integrating AI into digital forensics. Digital forensics can only be improved when AI can disclose its reasoning algorithm and discover large amounts of data in real-time. [21]

6.6 WAY FORWARD

6.6.1 Case Analysis of Cellebrite Guardian [22] for Privacy Management and Evidence Management

Cellebrite Guardian is a platform, which hosts a whole lot of evidence management tools focusing on digital forensic tools in collaboration with defense lawyers, investigators, and prosecutors for storing digital evidence on a portal placed on the Internet.

6.6.2 Workings of the Cellebrite Guardian

The investigator will log into the investigator portal and then submit all the devices to the forensic science laboratory. While making entry into the FSL, the investigating officer will have to give all the details related to the electronic device and make a request with respect to what the IO expects to be analyzed. The scientific officer at the forensic science laboratory will log into the platform and find the device's contents and attached legal documents (warrants, consent forms signed by the user, pictures of the device) and then build a report around the whole case and submit it to the investigating officer [23]. The following outlines how privacy management is done:

Making it privacy friendly by analysis of only relevant data: investigator will give all the data in the Cellebrite Guardian platform, then the examiner will download the device details to the Guardian worksheet, and the templates made for the forensic expert will allow the expert to fit in his findings into the predefined template given by the platform. All actions or data automated by Guardian Cellebrite will be in the form of trackable logs for maintaining the chain of custody. The forensic expert can preserve privacy by controlling who can access the forensic data. The platform runs a preliminary scan to get relevant data for the forensic expert, which is stored on the device under investigation through the native web review of the UFDR data, allowing the expert to evaluate the data from the extraction within the browser window itself, thus saving a lot of time for the investigator [24].

Classification of data extracted through a collaborative effort of the investigator, forensic expert, and investigating officer: There is limited access as to who can access the digital evidence, so it is possible that even the user of the device can access the contents of the forensic evidence through this web-based platform and help the forensic expert in classifying relevant private and non-relevant data to preserve his privacy [25].

6.6.3 Case Analysis of Veritone Redact Where Artificial Intelligence Is Being Used to Protect Privacy When Used in Digital Forensics

When a forensic investigation is underway, there is a need to protect the sensitive personal data of various individuals who are involved; this data includes Aadhaar numbers, credit card numbers, home addresses, email addresses, and passport numbers.

Veritone Redact works on the issue that manually redacting the personal information of the victim or user of the device is very time-consuming in light of privacy laws. To protect the privacy of the persons whose data is being handled by the investigator, manually protecting personal information is costly, time-consuming, and drives up the cost of the investigation [26].

What has Veritone done?

AI, privacy, and redaction software have been integrated to save time and costs. The AI software that has been made by Veritone for processing image, audio, and video evidence performs the following actions:

1. With one click, there can be redaction of heads and any objects that have been defined by the user in the image, video, and audio.
2. Automatic detection of objects like driver's license photos, laptop passwords, license plates, etc.
3. Displaying the detection in chronological order and that too within a player that is efficient at playing high-quality videos.
4. Where the user defines the objects to be private, the software will track the items throughout the video, and at the discretion of the investigating officer or forensic expert, they can be blurred or hidden.
5. Searching video and audio for keywords.
6. Redacting the sensitive personal data of individuals on the device.
7. Downloading logs of the redacted version, thus maintaining the chain of custody.
8. Collaborating and tracking the personal information that has been tracked by tagging the concerned files along with the updates on the status of the investigation that has been done.
9. Work can be performed by the forensic expert on the cloud or through a remote network; thus, the forensic expert can even work from home while assessing digital evidence. [27]

Beneficiaries:

1. Enforcement agencies
    1) Preservation of integrity of evidence
    2) Compliance with privacy law

    3) Compliance with procedures of the court
    4) Cost-effectiveness
    5) Less time-consuming
2. Public:
    1) Users can classify data that is private.
    2) Maintains integrity of investigation
    3) More trust and transparency
    4) Police save resources and time and can help the public in other areas.
    5) Identification of individuals is protected [28].

6.7 CONCLUSION

The development of AI-integrated digital forensic tools should follow ethical practices for privacy preservation by following the development standards and other industry standards that have been set by various agencies of the government. For example, the privacy standards that have been developed by the American Society of Crime Laboratory Directors for digital and multimedia forensics should be adhered to by AI for engaging in privacy-friendly digital forensics [29]. Enforcement agencies should use more tools like Cellebrite Guardian and Veritone Redact, which are privacy-friendly, cost-effective, and less time-consuming. AI software integrated with digital forensics should follow the following principles [30]. Knowledge representation ensures that by using standard domain ontology, AI can share all the information across and between forensic imaging tools; at the same time, a framework for discussion around the digital evidence can be enabled for the creation of a large and reusable case repository, thus helpful for the storage and retrieval of background knowledge. The reasoning process, whereby the reasoning algorithm that is used by AI lies within the domain of symbolic reasoning, has a decision-making process that has been shortened due to the absence of base rules, but this is majorly a disadvantage in digital forensics. To fix the reasoning algorithm, case-based reasoners ought to be used, which use past experiences in a familiar way, thus reducing the chances for errors to creep into the working of the forensic tool. Recognizing patterns, AI should be able to identify data clusters through pattern recognition and then classify data as per past experiences by running through nearby recognizers. To make this AI process successful, it is imperative to use AI-based neural networks and decision trees for identifying the initial patterns that are important in digital forensics. Data mining is performed by mining data through AI software, analyzing statistics, and assessing probabilities in quite an understanding manner, which may be used for pattern recognition, thus involving the process of exploration of data that majorly highlights the relation between information, the relationships of the data, and its consequent users; thus, AI should also depend on this process.

REFERENCES

[1] Dunsin, D., Ghanem, M. C., Ouazzane, K., & Vassilev, V. (2024). A comprehensive analysis of the role of artificial intelligence and machine learning in modern digital forensics and incident response. *Forensic Science International Digital Investigation*, *48*, 301675. https://doi.org/10.1016/j.fsidi.2023.301675

[2] How can forensic investigators gain an edge using AI? (2018, October 22). *Deloitte Switzerland*. https://www2.deloitte.com/ch/en/pages/forensics/articles/forensic-investigators-gain-an-edge-with-ai.html
[3] Faggella, D. (2019, February 2). AI for crime prevention and detection – 5 current applications. *Emerj Artificial Intelligence Research*. https://emerj.com/ai-sector-overviews/ai-crime-prevention-5-current-applications/
[4] Faruk, M. J. H., Shahriar, H., Valero, M., Barsha, F. L., Sobhan, S., Khan, M. A., Whitman, M., Cuzzocrea, A., Lo, D., Rahman, A., & Wu, F. (2021). Malware detection and prevention using artificial intelligence techniques. *2021 IEEE International Conference on Big Data* (Big Data). https://doi.org/10.1109/bigdata52589.2021.9671434
[5] Kietzmann, J., Paschen, J., & Treen, E. (2018). Artificial intelligence in advertising. *Journal of Advertising Research, 58*(3), 263–267. https://doi.org/10.2501/jar-2018-035
[6] Rughani, P. H. (2017). Artificial intelligence-based digital forensics framework. *International Journal of Advanced Research in Computer Science, 8*(8), 10–14. https://doi.org/10.26483/ijarcs.v8i8.4571
[7] Spillias, S., Tuohy, P., Andreotta, M., Annand-Jones, R., Boschetti, F., Cvitanovic, C., Duggan, J., Fulton, E. A., Karcher, D. B., Paris, C., Shellock, R., & Trebilco, R. (2024). Human-AI collaboration to identify literature for evidence synthesis. *Cell Reports Sustainability*, 100132. https://doi.org/10.1016/j.crsus.2024.100132.
[8] Naser, M. Z., & Alavi, A. H. (2021). Error metrics and performance fitness indicators for artificial intelligence and machine learning in engineering and sciences. *Architecture Structures and Construction, 3*(4), 499–517. https://doi.org/10.1007/s44150-021-00015-8
[9] Cioffi, R., Travaglioni, M., Piscitelli, G., Petrillo, A., & De Felice, F. (2020). Artificial intelligence and Machine learning applications in smart Production: Progress, trends, and directions. *Sustainability, 12*(2), 492. https://doi.org/10.3390/su12020492
[10] Jerian, M. (2023, September 6). Can AI be used for forensics and investigations? *Amped Blog*. https://blog.ampedsoftware.com/2021/10/05/can-ai-be-used-for-forensics-and-investigations
[11] Rigano, C. (2019). Using artificial intelligence to address criminal justice needs. *National Institute of Justice*. https://www.ojp.gov/pdffiles1/nij/252038.pdf
[12] Grimm, P. W., Grossman, M. R., & Cormack, G. V. (2021). Artificial intelligence as evidence. *Northwestern Pritzker School of Law Scholarly Commons*. https://scholarlycommons.law.northwestern.edu/njtip/vol19/iss1/2
[13] Lashkari, A. H., Li, B., Carrier, T. L., & Kaur, G. (2021). VolMemLyzer: Volatile memory analyzer for malware classification using feature engineering. *IEEE Access*. https://doi.org/10.1109/rdaaps48126.2021.9452028
[14] Iqbal, F., Debbabi, M., & Fung, B. C. M. (2020). Machine learning for authorship attribution and cyber forensics. In *International series on computer entertainment and media technology*. https://doi.org/10.1007/978-3-030-61675-5
[15] House, W. (2023, October 30). FACT SHEET: President Biden issues executive order on safe, secure, and trustworthy artificial intelligence. *The White House*. https://www.whitehouse.gov/briefing-room/statements-releases/2023/10/30/fact-sheet-president-biden-issues-executive-order-on-safe-secure-and-trustworthy-artificial-intelligence/
[16] Mökander, J., Juneja, P., Watson, D. S., & Floridi, L. (2022). The US algorithmic accountability act of 2022 vs. The EU artificial intelligence act: What can they learn from each other? *Minds and Machines, 32*(4), 751–758. https://doi.org/10.1007/s11023-022-09612-y
[17] Neuwirth, R. J. (2023). Prohibited artificial intelligence practices in the proposed EU artificial intelligence act (AIA). *Computer Law & Security Review, 48*, 105798. https://doi.org/10.1016/j.clsr.2023.105798

[18] Kumar, A., Shukla, P., Sharan, A., Mahindru, T., NITI Aayog, Sarkar, A., Nayan, A., & Asthana, K. (2018). National strategy for artificial intelligence. *National Strategy for Artificial Intelligence* [Report]. https://www.niti.gov.in/sites/default/files/2023–03/National-Strategy-for-Artificial-Intelligence.pdf
[19] Prasad-Rao, J., Fischer, O. J. P., Rowe, N. C., Williams, J., Puranik, T., Mavris, D., Natali, M., Tindall, M., Atkinson, B., Global Technology Connection, Inc., Aerospace Systems Design Laboratory, Naval Postgraduate School, & Naval Air Warfare Center Training Systems Division. (2023). Attrition risk and aircraft suitability prediction in U.S. Navy pilot training using machine learning. *Aerospace*, *1*, 0. https://faculty.nps.edu/ncrowe/TRAIT_paper_v5.pdf
[20] Ren, W., Wang, L., Choo, K-K. R., & Xhafa, F. (2019). *Security and privacy for big data, cloud computing and applications (computing and networks)*. Amazon.es: Libros. https://www.amazon.es/Security-Privacy-Computing-Applications-Networks/dp/1785617478
[21] Solanke, A. A., & Biasiotti, M. A. (2022). Digital Forensics AI: Evaluating, standardizing and optimizing digital evidence mining techniques. *KI – Künstliche Intelligenz*, *36*(2), 143–161. https://doi.org/10.1007/s13218-022-00763-9
[22] Revo4n. (n.d.). Managing evidence and investigations with Cellebrite guardian. *Revo 4n6*. https://revo4n6.com/blog-posts/f/managing-evidence-and-investigations-with-cellebrite-guardian
[23] Duc, H. N. (2023, October 31). Managing evidence and investigations with Cellebrite guardian. *eForensics*. Retrieved on December 2, 2025 from https://eforensicsmag.com/managing-evidence-and-investigations-with-cellebrite-guardian
[24] Kelly, L., Sachan, S., Ni, L., Almaghrabi, F., Allmendinger, R., & Chen, Y. (2020). Explainable artificial intelligence for digital forensics: Opportunities, challenges and a drug testing case study. In *IntechOpen eBooks*. https://doi.org/10.5772/intechopen.93310
[25] The challenges of collecting evidence. (2023, October 31). *eForensics*. Retrieved on December 2, 2024 from https://eforensicsmag.com/the-challenges-of-collecting-evidence/
[26] Veritone. (2024, June 10). How does AI protect privacy with redaction technology? *Veritone*. https://www.veritone.com/blog/how-does-ai-protect-privacy-with-redaction-technology/
[27] Wong, D. (2024, September 20). Everything about digital forensics & privacy. *Veritone Blog*. https://www.veritone.com/blog/digital-forensics-and-privacy/
[28] Kamble, D., & Salunke, M. D. (2024). Designing an automated, privacy-preserving, and efficient Digital Forensic Framework. *Journal of Autonomous Intelligence*, *7*(5), 1270. https://doi.org/10.32629/jai.v7i5.1270
[29] Firdonsyah, A., Purwanto, P., & Riadi, I. (2023). Framework for digital forensic ethical violations: A systematic literature review. *E3S Web of Conferences*, *448*, 01003. https://doi.org/10.1051/e3sconf/202344801003
[30] Harshita. (2023). Artificial intelligence in digital forensics. *Nasscom | The Official Community of Indian IT Industry*. https://community.nasscom.in/communities/emerging-tech/ai/artificial-intelligence-in-digital-forensics.html

7 Analyzing Vast Datasets with Machine Learning

Vineet Bhandari[1], Ganti Sri Rama Krishna[1], Chothmal Kumawat[1], and Narayan C. Debnath[2]

[1]School of Cyber Security and Digital Forensics, National Forensic Sciences University, Delhi Campus, New Delhi.

[2]Eastern International University, Vietnam.

7.1 INTRODUCTION

In the present interdependent world, tremendous growth in digital data has transformed the scope of digital forensics. Network devices and online parts of most investigations, but handling and analyzing such efforts are staggering tasks in need of tools and techniques that can process huge sums of data but give meaningful insights.

ML has turned out to be the much-needed solution in this sphere, providing the forensic investigator with automated and efficient methods of pattern discovery, anomaly detection, and classification of artifacts within the forensics domain. The strength of ML lies in teaching algorithms rather than relying on human-processed analysis like traditional approaches. This chapter will outline the role of ML in analyzing large dataset analysis in digital forensics, basic ideas regarding which ML techniques work [1], applications in real-world forensic contexts, and the problems associated with such data. The integration of ML into digital forensics would drastically enhance the capabilities of professionals in investigations and tackle problems of modern crime like cybercrime, fraud detection, and evidence classification.

We will thus discuss some of the foundational aspects of vast datasets and how ML contributes to data analysis. We will outline its impacts on solving contemporary forensic problems and then go over some of the challenges – ethical and practical ones – with a focus on scalable, interpretable, and transparent ML [2] solutions for digital forensics.

7.1.1 Understanding Vast Datasets in Digital Forensics

The dependence on digital systems in this current modern world has generated data that is unmatched to date, thereby propelling the field of digital forensics [3] by significant means and proportions. It encompasses a wide variety of structured and unstructured forms of massive datasets that are highly essential for forensic investigations [4], incident response in cybersecurity, and criminal evidence collection [5].

DOI: 10.1201/9781003512820-7

However, such data's enormous size and complexity raise huge challenges that demand new approaches for efficient handling and analysis.

7.1.2 Characteristics of Vast Datasets

Vast datasets in digital forensics can be described by the **three V's of big data** [6]:

- **Volume:** The huge amount of data coming from devices, cloud servers, network logs, and many other places. For instance, analyzing a one-year server log for a corporate network will easily result in terabytes of data and even make it inappropriate to be handled manually and demanding scalable solutions.
- **Velocity:** Data in forensic investigation is usually time-sensitive. For example, network logs and IoT device information are generated at tremendous speeds. They may need immediate analysis to prevent or respond to cybersecurity breaches.
- **Variety:** Data for digital forensics can be in several formats, including text (email, logs), multimedia (images and videos), binary files, and metadata. All these forms require preprocessing to standardize and process for analysis.

Additionally, large amounts of data to process are often very noisy, redundant, and irrelevant. For instance, probably only a small fraction of network traffic logs could be applicable for detecting intrusion attempts, while the rest adds overhead to the processing pipeline.

7.1.3 Sources of Vast Datasets

Digital forensic datasets originate from multiple sources, each presenting unique challenges:

- **Network Traffic Logs:** Network infrastructure generates traffic logs and intrusion detection system (IDS) outputs, which are critical for identifying unauthorized access or unusual activities.
- **Cloud Services:** With the growing adoption of cloud computing, digital evidence is often stored across multiple servers. Forensic analysis of cloud data introduces complexities such as encryption and jurisdictional constraints.
- **Social Media Platforms:** Social media data is rich in behavioral insights but can be ephemeral (e.g., disappearing messages) or protected by privacy measures.
- **IoT and Wearable Devices:** IoT systems generate continuous streams of data. Examples include health data from wearables or environmental data from smart home systems, which can be pivotal in criminal cases.
- **Mobile Devices:** Mobile devices are primary sources of evidence, encompassing app logs, geolocation data, and communication records. With encrypted storage and advanced security mechanisms, accessing and analyzing this data is a challenge.

7.1.4 Challenges in Handling Vast Datasets

- **Scalability:** Most of the traditional forensic tools are not designed to handle large-scale computations. Distributed computing frameworks like Apache Hadoop and Apache Spark are increasingly coming into use to scale up on the processing capability.
- **Data Quality:** Large datasets often contain incomplete, inconsistent, or redundant data. Preprocessing methods such as deduplication and data normalization are essential but time-consuming.
- **Privacy and Ethical Concerns:** Accessing and analyzing personal data for forensic purposes often intersects with privacy laws, like GDPR and CCPA. Adherence to ethical standards while performing investigations remains a critical concern.
- **Storage and Retrieval:** Managing the storage of vast datasets over time and ensuring efficient retrieval for investigations adds operational overhead.

7.1.5 Role of Machine Learning in Managing Vast Datasets

Machine learning has been particularly useful in data analysis and processing very large data [7]. It automates labor-intensive activities in feature extraction, anomaly detection, and data classification, enabling forensic investigators to devote themselves to the most significant findings, as follows:

- **Feature Selection:** ML algorithms are designed to select the most relevant features in a dataset, reducing dimensionality and improving analysis speed.
- **Anomaly Detection:** This includes unsupervised learning models, such as k-means clustering and Autoencoders, used to detect anomalies or outliers in data.
- **Automation of Evidence Classification:** ML models classify data into classifications including harmful actions, typical conduct, and possible threats.

The positive side of digital forensics lies in the large volumes of data—it offers. More opportunities, as well as more substantial evidence and valuable findings. Actually, large collections of data pose high demands on sophisticated tools and methods for correct data management and analysis, such as machine learning. More challenges will come with developing forensic science.

7.2 MACHINE LEARNING FUNDAMENTALS FOR DATA ANALYSIS

Machine learning, a subset of AI, is the set of algorithms designed and used to build predictive models directly from data with as little explicit programming as necessary. This is one field where ML has exploded exponentially across a broad spectrum of disciplines in the last few decades and has become an inalienable tool for data analysis, helping to spot patterns, formulate predictions, and derive actionable insights from large and complex datasets. The basic axiom of ML is the ability to learn and improve performance iteratively with the assimilation of increasing quantities of data so that there is no requirement to have hand-crafted rules and instructions specific to a particular task. As data volumes continue to expand, ML systems adapt, enhancing their predictive accuracy and robustness by leveraging the inherent structure within the data, which allows for the formulation of sophisticated models capable of making generalizable predictions across diverse contexts.

7.2.1 Overview of Machine Learning

Machine learning may be broadly categorized into three paradigms: supervised learning, unsupervised learning, and reinforcement learning [8]. These categories pose distinct approaches for training algorithms and solving various data analysis problems.

7.2.2 Supervised Learning

Supervised learning is one of the most basic machine learning approaches, mainly used for data analysis. In the supervised learning paradigm, algorithms learn from labeled datasets but input features x are paired with their respective targets y. The goal is to find a function f that maps x to y based on minimal error to make predictions on unseen data [9]. It is therefore mathematically represented as

$$y = f(\mathbf{x}) + \epsilon \tag{7.1}$$

where ϵ represents the error term accounting for deviations from true values. To optimize the model, the difference between predicted (y) and actual ($\hat{y}$). Its goal is to be minimized through loss functions such as MSE, given actual values ($\hat{y}$). Basic algorithms in supervised learning include linear regression, decision trees, and support vector machines (SVM). For example, linear regression defines the relationship between the feature vector and the target variable as follows:

$$\hat{y} = \mathbf{w}^T \mathbf{x} + b \tag{7.2}$$

where $\mathbf{w}$ stands for feature weights, $\mathbf{x}$ is the input vector, and b is the bias term. Applications span from medicine predicting health outcomes to finance and beyond forecasting. With the ability to use labeled data, supervised learning is essential for many of the productive domains.

7.2.3 Unsupervised Learning

In unsupervised learning, datasets lack output labels, and the goal is to uncover hidden structures or patterns within the data. Common applications include clustering, dimensionality reduction, and anomaly detection, focusing on extracting insights rather than predicting target variables [10].

A very popular method is k-means clustering, which partitions data into k clusters by minimizing the within-cluster variance. The objective function is

$$J = \sum_{i=1}^{k} \sum_{x_j \in C_i} \| x_j - \mu_i \|^2 \| \tag{7.3}$$

where C_i are points in the cluster i, μ_i is the cluster centroid, and x_j μ_i^2 are the squared distances from a point x_j to μ_i. The algorithm iteratively updates the clusters to minimize the bad compactness and separation.

The role of unsupervised learning is crucial in domains such as **healthcare** and **sensor networks**. For example, clustering techniques organize patients exhibiting analogous conditions to facilitate accurate diagnoses, whereas anomaly detection serves to identify irregularities, including system malfunctions or security vulnerabilities.

7.2.4 Reinforcement Learning (RL)

RL is the paradigm of machine learning where an agent learns how to make decisions from environmental interactions, taking rewards or penalties as feedback. Unlike supervised learning, RL does not rely on any labeled data; rather, an optimal policy π mapping states s to actions a needs to be developed to maximize cumulative rewards over time [11].

The expected long-term reward that may be accumulated when starting from a given state s is estimated by the value function $V(s)$. The action-value function $Q(s, a)$ on the other hand expresses the expected return that may arise after executing an action in state s. Such functions obey the Bellman equation:

$$V(s) = \mathbf{E}[R_t + \gamma V(s_{t+1}) \mid s_t = s] \tag{7.4}$$

where R_t is the reward obtained at time t, γ is the discount factor that is applied to future rewards, and s_{t+1} is the next state achieved. This equation enables iterative updates in hopes of improving the policy.

RL has far-flung applications. Robotics learns these policies to autonomously interact with the environment, while in gaming, it optimizes strategies for non-player characters. In healthcare, RL provides support for personalized treatment planning and handling uncertainties in medical decision-making [12].

7.2.5 Core Concepts in Machine Learning for Data Analysis

Several foundational concepts are integral to understanding machine learning algorithms and their applications in data analysis [13].

7.2.5.1 Features and Labels

In machine learning, datasets are usually represented by features and labels. Features are input variables that the machine learning model uses to make predictions, while labels represent the target output the model aims to predict. For example, in a predictive model for house prices, the features could include the number of bedrooms, square footage, and location, while the label would correspond to the house price [14].

7.2.5.2 Training and Test Data

In training and testing machine learning algorithms, it is common practice to split the available data into two separate parts: training and testing. The training process is used to adapt the model to learn basic patterns and relationships across objects, while the testing process is used to evaluate the ability of the model to extend to previously unseen objects. From this distribution, the performance of the model can be inferred by adapting it to the training data without overtraining. This increases the repeatability of the evaluation model by reducing the dependency on specific, stable data. In k-fold cross-validation, the dataset is split into subsets or folds of approximately equal size. In *k* train time, in each training session *k* 1 folds are used for training and the remaining one fold is used for testing. This will provide an opportunity to evaluate the performance of each data source, as each data point will appear only once in training and testing. The final performance measure is usually the average of the number of runs up to k to provide a better estimate of the standard error.

7.2.5.3 Overfitting and Underfitting

A major problem facing machine learning studies is how to strike the optimal balance between overfitting and underfitting. Overfitting occurs when the model becomes too complex and introduces noise or variability into the training data. This results in a model that fits well to training but is generally poor at what is not seen, thus weakening the accuracy of new cases. Generalization is ineffective when it is faulty; this occurs when the model is too simple to capture the underlying structure and complexity present in the data [15]. The effectiveness of both training programs and training methods is not ideal.

To avoid overfitting, regularization methods are often used. Regularization imposes some additional constraints on the model to prevent it from becoming too complex, thus improving its generality. A very popular technique, especially in linear regression, is *L*2 normalization. The *L*2 objective function constant and the time penalty can be written as:

$$J_{reg} = J(w) + \lambda \sum_{i=1}^{n} w_i^2 \tag{7.5}$$

This means that, without further modification, λ is used to represent a constant that controls the balance between complex models and fits to the data shown, while $J(\mathbf{w})$ represents the initial cost function, which is ignored. Adding the term λ L*n* 2 penalizes large model coefficients w_i, thus avoiding overly complex models and encouraging smoother, more complex methods of problem solving.

7.2.6 Basic Types of Machine Learning Algorithms

Machine learning algorithms can be categorized into several types based on the approach they take. Below, we outline the basic categories of algorithms commonly used in data analysis.

7.2.6.1 Linear Regression

Linear regression is one of the most widely used algorithms in supervised learning for predictive modeling. The goal of linear regression is to establish a relationship between the input feature vector **x** and the target variable *y* so that it can predict *y* from the new value of the file. Mathematically, the relationship between the input and target variables can be calculated from the following equation:

$$y = \mathbf{w}^T \mathbf{x} + b, \tag{7.6}$$

where **w** $\mathbf{R}^n$ is the weight vector related to the input feature **x** and b **R** is the bias term. The weight vector **w** and the time bias b are the parameters that the model will learn from the training process. The squared difference between estimated values $\hat{y}$ and the true target y serves as the loss function, which the model seeks to minimize during training to improve prediction accuracy. This is done by optimizing the following values:

$$J(\mathrm{w}, b) = \frac{1}{m} \sum_{i=1}^{m} (\hat{y}_i - y_i)^2, \tag{7.7}$$

where m represents the training examples, $\hat{y}_i$ is the model predicted value of the ith example, and y_i is the true value. Here, the goal is to find the values between **w** and b that minimize the function $J(\mathbf{w}, b)$.

7.2.6.2 Logistic Regression

Logistic regression is one of the simple methods used in educational monitoring to implement binary distribution functions. It aims to predict one of two outcomes labeled as categories 0 and 1, given the input feature vector **x**. Unlike regression, logistic regression uses a logistic or sigmoid function applied to a combination of input variables [16] to model the probability that the target variable y belongs to a particular group, such as category 1 (e.g., $y = 1$). The probability of observing class 1 is given by:

$$P(y = 1 \mid X) = \frac{1}{1 + e^{-(w^T X + b)}} \tag{7.8}$$

where w $\in R^n$ represents the body of the weight vector effect, the input features x and b $\in R^n$ are the deviation points. The sigmoid function $\sigma(z) = \frac{1}{1 + e^{-(w^T \mathrm{X} + \mathrm{b})}}$ maps the linear output $\mathbf{w}^T \mathbf{x} + b$ to the probability value which is in the interval [0, 1], i.e., the probability of the sample being in category 1. This is the difference between the predicted probability and the true binary label. The binary cross entropy loss function is given by

$$L(\mathrm{w}, \mathrm{b}) = -\frac{1}{m} \sum_{i=1}^{m} [\mathrm{y}i \log P(\mathrm{y}_i = 1 \mid \mathrm{x}_i) + (1 - y_i) \log(1 - P(yi = 1 \mid \mathrm{x}_i))] \tag{7.9}$$

where m is the number of training examples, y_i is the true binary label of i for the first example, and $P(y_i = 1\mathbf{x}_i)$ is the estimated probability of category 1 for the ith example. The goal of the optimization process is to find the parameters **w** and b that satisfy the minimum loss function, so that from the parameters, a good model is designed for classifying new ideas.

7.2.7 Decision Trees

Decision trees are non-linear models that can be applied for classification and regression tasks. The algorithm splits data into subsets based on input features with the objective of maximizing information gain (in the case of classification) or minimizing variance (in regression). A decision tree structure can be represented as a tree in which each internal node represents a decision based upon a feature, and each leaf node represents a prediction [17].

7.2.7.1 Support Vector Machines (SVMs)

SVM is a type of supervised learning algorithm that is used only for classification. The main idea behind SVM is to find a separating hyperplane that allows different data points to be separated. An edge is defined as the distance between the hyperplane and the nearest data point on either side (called the support vector). Mathematically, the decision function in SVM [18] is shown as

$$f(\mathbf{x}) = \mathbf{w}^T\mathbf{x} + b \tag{7.10}$$

The weight vector **w** is perpendicular to the hyperplane. The input feature vector is **x** and the bias term is b. The goal of SVM is to maximize the size of the training classifier such that all the training data are correctly classified. Mathematically, this can be formulated as the following optimization problem:

$$y_i(\mathbf{w}^T\mathbf{x}_i + b) \geq 1, \forall i \tag{7.11}$$

where y_i 1, 1 represents the category name of the ith entry and $\mathbf{x}_i$ is the character vector of this entry. Optimization problems are usually solved using quadratic programming methods to find **w** and b such that the pair (**w**, b) best describe the decision space. Kernel method: When the data is not linearly separable, the kernel method is used to implicitly represent the data in a higher feature space. This makes it easy to use linear hyperplanes to separate the data.

7.2.7.2 K-Nearest Neighbors (KNNs)

The KNN algorithm is a simple yet effective nonparametric method for classification and regression tasks. The function of KNN is to determine the nearest neighbors of the problem and then make predictions based on the majority of the class (in the case of classification) or the average of the neighbors (in the case of regression) [19]. More precisely, given the new data point **x**, the prediction can be written as

$$\hat{y} = \frac{1}{k}\sum_{i=1}^{k} y_i \tag{7.12}$$

where y_i corresponds to the class label (in classification) or target variable (in regression) of the ith nearest neighbor **x**. The distance metric used to calculate closeness is usually the

Euclidean distance, but there are other distance metrics such as the Manhattan distance or the Minkowski distance. Instead, the model stores all the training data and makes predictions by determining the closest point during inference.

7.2.7.3 Machine Learning for Data Analysis

Machine learning represents a very strong tool to analyze complex datasets and make predictions accurately. Data scientists can leverage a variety of algorithms, such as SVM, KNN, and others, to identify potential hidden patterns, classify data, and help forecast future trends in various domains. These successful algorithms enable drawing actionable conclusions, which act as driving forces in leading improvements in areas such as healthcare, finance, and engineering. With the pace of advancement in machine learning, its integration into data analysis workflows will definitely accelerate innovation and improve decision-making capabilities across diverse industries.

7.3 DATA VISUALIZATION IN MACHINE LEARNING

Visualization performs two roles in machine learning: understanding data and understanding model behavior. Visualization of data becomes an exploratory pre-modeling technique for investigating datasets, uncovering patterns, trends, or anomalies that have effects on downstream tasks. Post-modeling, visualization facilitates the interpretation of predictions, model performance, and even helps in debugging [20]. This article delves into the techniques and tools, as well as some theoretical underpinnings, that make visualization an integral part of any machine learning workflow.

7.3.1 Data Visualization in the Pre-Modeling Phase

In the pre-modeling phase, it takes understanding and structuring of the given dataset for further training and evaluation. With the help of diversified visualization techniques and graphical representation, the structure of the data can be explored [21].

7.3.1.1 Descriptive Visualizations

- **Scatterplots**: Represent the relationship between two continuous variables. Consider two continuous features X_1 and X_2: scatterplots can help represent their relationship:

$$r = \frac{\sum_{i=1}^{n}(X_1 - \bar{X}_1)(X_2 - \bar{X}_2)}{\sqrt{\sum_{i=1}^{n}(X_1 - \bar{X}_1)^2 \sum_{i=1}^{n}(X_2 - \bar{X}_2)^2}} \tag{7.13}$$

  where r measures the Pearson correlation coefficient [22].
- **Histograms**: The frequency distribution of a single characteristic can be summarized and facilitate the understanding of skewness and outliers. The histogram can be represented as:

$$f(x) = \frac{\text{Number of values in bin}}{\text{Total number of values}}.100 \tag{7.14}$$

  where $f(x)$ denotes the percentage of occurrences falling within each interval [23].

- **Box Plots (Box-and-Whisker Plots)**: Display the distribution of data, emphasizing some of the important statistical quantities, which are the median, quartiles, and possible outliers. The interquartile range is defined to be

$$IQR = Q_3 - Q_1 \tag{7.15}$$

where Q_3 is the upper quartile (75th percentile) and Q_1 is the lower quartile (25th percentile). Typically, outliers are all points below $Q_1 - 1.5 \times IQR$ or above $Q_3 + 1.5 \times IQR$.

- **Bar Charts**: Used primarily for comparing quantities among different categories, the height or length of each bar indicates the value related to its category.

$$\text{Height of bar} = \frac{\text{Category value}}{\text{Total sum of all categories}} \times 100 \tag{7.16}$$

This technique is used to show how each category compares as a percentage of the total.

- **Heatmaps**: Present the information in a matrix layout, such that the cell intensity shows its associated value. The value inside a heatmap A_{ij} can be displayed in the form of

$$A_{ij} = \textit{value of the element at row}\, i\, \textit{and column}\, j \tag{7.17}$$

The intensity of the color is actually the magnitude of each value, making it easier to spot patterns [24].

- **Line Charts**: Illustrate patterns across time or sequential data points. The formula for a line may be expressed as

$$y = mx + b \tag{7.18}$$

where y is the dependent variable, x is the independent variable, m represents the slope, and b is the y-intercept.

- **Violin Plots**: A combination of a box plot and density plot: the width of the "violin" at a certain feature value equals the approximate density at that point. The kernel density estimation formula is given as

$$f(x) = \frac{1}{nh} \sum_{i=1}^{n} k\left(\frac{x - x_i}{h}\right) \tag{7.19}$$

where $f(x)$ is the estimated density, n is the number of data points, h is the bandwidth, and K is the kernel function, mostly Gaussian [25].

- **Pair Plots (Scatter plot matrix)**: Show pairwise relationships between multiple variables. If there are n features, a pair plot will show $n(n-1)$ scatter plots of all combinations of features.

- **Pie Charts**: Illustrate as fractions or percentiles of the whole. Each of the wedges has a measure that represents the percent of each category:

$$\theta_i = \frac{X_i}{\sum_{i=1}^{n} X_i} \times 360^{\circ} \tag{7.20}$$

where θ_i is the angle corresponding to the ith slice and X_i is the value associated with the ith category, while summation aggregates all categories.

- **Radial Charts (Spider Plots)**: Represent data that contain more than one variable using a circular arrangement. The data points representing each axis are connected to form a polygonal shape.

$$r_i = value\ of\ the\ feature\ for\ the\ i^{th}\ axis \tag{7.21}$$

where each axis represents a specific characteristic, and the points are joined to form a radial structure.

- **Density Plots**: Present the probability density function associated with a continuous variable. The equation used to express a kernel density estimate (KDE) is the same as given in (7.19).
- **Stacked Area Charts**: Used when a number of categories are accumulated and depicted on the same graph over time or another continuous variable. Cumulative value is plotted as:

$$yt = \sum_{i=1}^{n} yit \tag{7.22}$$

where y_t represents the cumulative value at time t and y_{it} the value of the ith category at time t.

- **Dot Plots**: Individual data points are displayed along one axis. For the data points labeled as $x_1, x_2, \ldots, x_n$, the dot plot can be drawn as follows:

$$\text{Dot at position } x_i \text{ for each } i \in \{1, 2, \ldots, n\}. \tag{7.23}$$

In this context, each point symbolizes an individual observation within the dataset.

- **Treemaps**: Represent hierarchical data using nested rectangles. Each rectangle's area is proportional to the value it represents:

$$\text{Area of rectangle} = \frac{X_i}{\sum_{i=1}^{n} X_i} \tag{7.24}$$

where X_i indicates the value assigned to the ith measurement.

- **Bubble Charts**: A scatter plot variation whose data points are represented as bubbles. The area of the bubble represents an additional dimension:

$$A_i = \pi r_i^2 \tag{7.25}$$

where r_i is the radius of the bubble and the value related to the area A_i.

7.3.1.2 Dimensionality Reduction for Visualization

So, high-dimensional data are very often reduced, saving only the intrinsic structure in those data:

- **Principal Component Analysis (PCA)**: Projects data to a lower-dimensional subspace as $Z = XW$, where X represents the data matrix, W carries the eigenvectors of the covariance matrix, and Z represents the transformed data in the reduced dimensional space.
- **t-Distributed Stochastic Neighbor Embedding (t-SNE)**: Maps high-dimensional data to a 2D or 3D space, preserving neighborhood relationships. The similarity between points i and j in t-SNE is modeled as

$$p_{ij} = \frac{\exp(-\| x_i - x_j \|^2 / 2\sigma^2)}{\sum_{k \neq l} \exp(-\| x_k - x_l \|^2 / 2\sigma^2)} \tag{7.26}$$

  where the hyperparameter σ is adapted according to local density. It is a non-linear dimensionality reduction technique aiming to minimize the Kullback-Leibler (KL) divergence between the joint probability distributions in high-dimensional space and the low-dimensional embedding:

$$KL(P \| Q) = \sum_{i,j} P_{ij} \log \frac{p_{ij}}{Q_{ij}} \tag{7.27}$$

  where P_{ij} is the similarity measure between the high-dimensional space, while Q_{ij} is the similarity measure between the low-dimensional space.
- **Linear Discriminant Analysis (LDA)**: A supervised projection technique of reducing the dimensionality of data to maximize class separability. The projection
  matrix W is found as

$$W = \arg\max_{w} \frac{\det(S_B)}{\det(S_W)} \tag{7.28}$$

  where S_B is the between-class scatter matrix and S_W is the within-class scatter matrix. The goal is to maximize the ratio of between-class variance to within-class variance.
- **Autoencoders**: It is a kind of neural network that can apply unsupervised dimensionality reduction. The encoder maps the input x to a lower-dimensional space z, and the decoder tries to reconstruct the input:

$$z = f(x, \theta_e), x = \hat{g}(z, \theta_d) \tag{7.29}$$

  where f is the encoder function; g is the decoder function; θ_e and θ_d are the weight parameters for the encoder and decoder, respectively; and $\hat{x}$ is the reconstructed input.

- **Independent Component Analysis (ICA)**: A technique to decompose a multivariate signal into additive statistically independent components. ICA tries to find a matrix W such that the transformed data Z are statistically independent:

$$Z = XW \tag{7.30}$$

In this context, X is the observed dataset and W is a matrix applied to transform the data space into independent components space. The independence among them is often measured using some metric, such as negentropy.

- **Isomap**: A non-linear dimensionality reduction technique that relies on multidimensional scaling (MDS) is used. This algorithm tries to preserve geodesic distances between the data points. First, the algorithm constructs a neighborhood graph and subsequently finds the shortest path between pairs of points. The ultimate goal of the algorithm is to solve the following optimization problem:

$$\min_{Y} \sum_{i,j} (d_{ij} - \| y_i - y_j \|)^2 \tag{7.31}$$

where d_{ij} denotes the geodesic distance between the high-dimensional points i and j in this context, while y_i and y_j are points from the respective low-dimensional space.

- **Uniform Manifold Approximation and Projection (UMAP)**: A non-linear dimensionality reduction technique similar to t-SNE but faster and more scalable. UMAP preserves both local and global structures. It is an optimization of the best low-dimensional projection minimizing a trade-off cost function between preserving local and global structure:

$$\text{minimize} \sum_{i,j} \| y_i - y_j \|^2 \tag{7.32}$$

In this context, y_i and y_j denote the embeddings in the reduced-dimensional space, with distances being computed using a fuzzy simplicial set representation of the data.

- **Factor Analysis**: A statistical method used for modeling observed variables as linear combinations of potential unobserved variables – also known as factors. The observed data matrix X can be written as follows:

$$X = LF + \epsilon \tag{7.33}$$

where L is the matrix of factor loadings, F is the matrix of factors, and ϵ is the error term. Factor analysis is required to identify the underlying factors, which would actually explain the observed correlation.

- **Non-Negative Matrix Factorization (NMF)**: A matrix factorization method that decomposes a matrix V into two nonnegative matrices, W and H:

$$V \approx W\ H \tag{7.34}$$

In that regard, W is known as the basis matrix and H as the coefficient matrix. NMF has been proven to work very effectively in applications where the involved data is non-negative, including image processing and mining text.

- **Multidimensional Scaling (MDS)**: A technique used for the reduction of data dimensionality, retaining pairwise distances between data points. MDS minimizes the stress function:

$$\text{Stress}=\sqrt{\sum_{i,j}(d_{ij}-\| y_i - y_i \|)^2} \tag{7.35}$$

where d_{ij} is the pairwise distance in the high-dimensional space and y_i, y_j are the corresponding points in the low-dimensional space. MDS aims to find the configuration of points in the low-dimensional space that best preserves the pairwise distances from the original high-dimensional space.

7.3.2 Explainability and Interpretability through Visualization

(i) **Feature Importance**: Feature importance is the process of finding which features in the dataset are highly influential on the decisions made by a predictive model. This is predominantly used with tree-based models, such as Random Forest and XGBoost. The importance of a feature is essentially summed at each node in the tree structure. The formula for feature importance is

$$\text{Importance}(f)=\sum_{t\in T} I_t(f) \tag{7.36}$$

where f is the feature, t is a node in the tree, and $I_t(f)$ is the importance of feature f at node t. Feature importance helps in identifying critical features for making predictions and reducing dimensionality by choosing only the most important features.

(ii) **SHapley Additive exPlanations (SHAP) Values**: SHAP values, based on game theory, describe individual predictions by associating a "fair" contribution to each feature. The formula to calculate the SHAP value of feature i is

$$\phi_i = \sum_{S\subseteq F\setminus\{i\}} \frac{|s|!(|F|-|S|-1)!}{|F|!}\left[f(S\cup\{i\})-f(S)\right], \tag{7.37}$$

where ϕ_i is the value assigned to the feature i and S is a set of all features except for i. Besides, F denotes a universal set of features and $f(S)$ is the model's output predicting assuming set S. SHAP values are explanatory of the black-box models, and it can even make it possible for one to visually see which feature has a positive or a negative effect on the overall prediction.

(iii) **Gradient-Weighted Class Activation Mapping (Grad-CAM)**: It is a visualization technique used along with convolutional neural networks (CNNs), which improves the attention regions in the input images responsible for influencing the model's prediction. The Grad-CAM computation for the class c is calculated as follows:

$$L^{c}_{Grad-\mathrm{CAM}} = \mathrm{Re\,LU}\left(\sum_{K} \alpha_k^c A_k\right) \tag{7.38}$$

where $\alpha^c =^{\partial y} c$ denotes the gradients computed weights and A_k denotes feature maps generated by the k-th layer. Grad-CAM proves to be a very helpful visualization aid to show how CNNs interpret specific regions in an image for a particular classification task, and it is useful for debugging the CNN to make sure that the network is focusing on relevant features.

(iv) **(Local Interpretable Model-Agnostic Explanations (LIME)**: It locally approximates complex models around a specific prediction by building a simpler, more interpretable model, for example, linear regression based on the predictions generated by perturbed samples. This calls for generating perturbations of the input instance, producing predictions on the perturbed samples by the use of the black-box model, and then fitting a simpler model to the localized predictions. LIME can be used to explain a black-box model's predictions on specific instances and has been applied to a variety of fields, including tabular data, text, and images.

(v) **Attention Maps**: Attention mechanisms point to the input parts on which a model makes predictions. It is a very common technique in NLP as well as in vision tasks. For the transformer, the computation for its attention mechanism is as follows:

$$\mathrm{Attention}(Q,K,V) = \mathrm{softmax}\left(\frac{QK^T}{\sqrt{d_k}}\right)V \tag{7.39}$$

where Q is the query matrix, K is the key matrix, V is the value matrix, and d_k is the dimension of the key/query vectors. Attention maps are helpful for understanding which words within a sentence are emphasized in NLP tasks, as well as pointing to relevant regions within an image that are relevant in computer vision tasks.

(vi) **Decision Tree Visualization**: For interpretable models like decision trees, visualizing the tree structure itself helps explain how predictions are made. The features are split via the values of the feature, and each path from the root to a leaf represents a decision rule. This can easily be visualized with decision trees for making decisions and staying on the logic that is employed in making the prediction [26].

(vii) **Surrogate Models**: These are known as interpretable models. They include decision trees and linear models, trained to approximate the predictions of a more complex black-box model, so as to understand what the complex model is doing overall. Then, by visualizing the surrogate model's decision boundaries or important features, one can interpret without directly accessing the complex model.

(viii) **Global Sensitivity Analysis**: It considers the degree to which variations in the input variables of a model impact the output over the entire dataset. This type of analysis facilitates the discovery of the most influential features in predictions and can be presented using sensitivity plots or heatmaps that show the impact of each feature in the feature space. The calculation for the sensitivity of every feature can be compared as

$$S_i = \frac{1}{N}\sum_{k=1}^{N} | \frac{\partial yk}{\partial xi} | \tag{7.40}$$

where S_i is the sensitivity of feature x_i with respect to the model's output and N is the number of samples in the dataset.

(ix) **Class Activation Maps (CAMs)**: CAMs are similar to Grad-CAM but work specifically for fully connected layers or less complex models. They highlight the regions of an image that contribute most to the classification decision. CAMs are created by using the weights of the final layer to produce a weighted sum of feature maps, helping interpret CNN-based model predictions. The CAM is computed as

$$\text{CAM} = \sum_{k} \alpha_k \cdot A_k \tag{7.41}$$

In this respect, α_k is the weight of the last fully connected layer and A_k is the feature maps created from the final convolutional layer. The resulting weighted summation produces a heatmap that highlights regions within the image that are most influentially impacted by its classification outcome.

7.3.3 Challenges and Future Directions in Data Visualization

- **High-dimensional data**: Generally, high-dimensional data refers to sets of data that contain many features or variables, so that the "curse of dimensionality" occurs. The more dimensions involved, the sparser and thus harder the data becomes to visualize well. High-dimensional datasets are common in applications involving genomics or image processing, requiring techniques such as dimensionality reduction (PCA, t-SNE) on such datasets so the data can be interpretable.
- **Selecting appropriate visualization techniques**: Choosing a proper technique for visualization may be of great importance to the effective transfer of data insights. The choice of a type of visualization depends on factors such as the type of data (discrete and continuous), the dimensionalities involved, and objectives of analysis. For example, bar charts are very good for comparing categorical values, whereas heatmaps are more useful to represent relationships between variables visually.
- **Interpretability of complex visualizations**: Complex visualizations, such as multidimensional plots or network graphs, can be difficult to interpret, especially for non-experts. To ensure interpretability, it's important to

design visualizations with clear labels, intuitive color schemes, and interactive elements that allow users to explore the data further, simplifying the key insights and patterns.

- **Domain-specific visualization techniques**: Complex visualizations, whether multidimensional plots or even network diagrams, are hard to interpret – even by an expert – for those unfamiliar with them. The creation of those visualizations must be accompanied by user-friendly labeling and intuitive color use as well as interactive functionality that permits the focus of further points of investigation that reveal key insights and patterns.
- **Ethical considerations**: Ethical practices in data visualization ensure the facts are represented accurately without scaling or distortion. Therefore, there is no misrepresentation, adjustment of scales on axes that prove inconvenient, and confidential information remains secret and safe. Ethical methodologies in practice allow for transparency regarding the findings resulting from the application of the visualization practices.
- **Emerging trends**: New trends in data visualization keep emerging, such as interactive visualization, like the use of AR/VR, and possibly the use of AI and machine learning to further enhance or automate visualizations. Other new ways of showing real-time data also emerge. This keeps the visualizations up to date with new technologies so that improvements can be made in analysis and user experience.

Machine learning and data visualization are highly interconnected as the former can be utilized to gain deeper insight into the properties and behaviors of models. Applying applied mathematics along with intuitive graphical models and up-to-date technological elements in visualization can aid practitioners in drawing usable conclusions, maximizing model performance, and enhancing interpretability. This interdependency of machine learning and visualization fosters superior and more informed decision-making across various domains, apart from promoting understanding.

7.4 TYPES OF ANALYSIS USING MACHINE LEARNING

7.4.1 Classification Analysis

Classification refers to the task of statistical categorization, where it classifies data under known predefined classes by following patterns obtained from labeled datasets. It is a supervised learning algorithm that learns to create a mapping from known data with known outcomes towards building a model to generalize on new, unseen examples. Classification mathematically tries to approximate a function $f : X \rightarrow Y$, where X denotes the input features and Y is the target variable.

For example, decision trees classify a dataset based on some cut-off, whereas SVM tries to optimize the placement of the separating hyperplane between classes. Neural networks are multilayer architectures representing very complex nonlinear patterns, and logistic regression models the probabilistic outcome based on the sigmoid function.

$$P(y=1 \mid X) = \frac{1}{1+e^{-(\beta_0+\beta_1 X_1+\ldots+\beta_P X_P)}} \tag{7.42}$$

Medical diagnoses, for example, involve algorithms that come up with a decision as to whether the tumor is malignant or benign given its size and texture. Intrusion detection systems classify network traffic as either benign or malicious depending on parameters analyzed, including packet size and transmission rates.

7.4.2 Clustering Analysis

Clustering: partitions unlabeled data into groups based upon the inherent similarities. It is unsupervised learning where the goal is to minimize intra-cluster variance while maximizing inter-cluster separation:

$$\arg\min_s \sum_{i=1}^{k} \sum_{x \in S_i} \| x - \mu_i \|^2, \tag{7.43}$$

where S_i symbolizes the clusters and μ_i are their corresponding centroids. The methods that are most typically used include K-means, which, in turn, allocates points systematically into clusters depending on the distances to the centroids, hierarchical clustering, creating a hierarchy, and Gaussian mixture models, which infers data as a mixture of Gaussians. Applications include customer segmentation in a marketing context, in which people are segmented based on behaviors and preferences for specific targeted marketing campaigns. For example, IoT data in smart cities could be clustered to improve resource allocation and manage urban infrastructure better.

7.4.3 Anomaly Detection

Anomaly detection typically highlights patterns that do not fit into established norms and can represent the most significant events, such as fraud or system failures. This type of analytical method is essential in monitoring dynamic systems for fault detection. Techniques include one-class SVM delineation of normal data, autoencoders reconstructing input data, and quantifying the deviation as potentially an anomaly. Statistical models express the distributions of the data, and techniques such as isolation forest divide the data space to find anomalies. In the field of cybersecurity, anomaly detection is crucial in identifying uncharacteristic patterns associated with network traffic, which could be indicative of a DDoS attack, among other cyber threats. The integration of machine learning enhances intrusion detection systems' effectiveness by allowing examination of high-dimensional data for subtle anomalies that may be indicative of malicious activities.

7.4.4 Regression Analysis

Regression provides the relationship between the dependent and independent variables with respect to continuous outcomes. The aims of regression analysis are to minimize the prediction error. The expressions are usually

$$\arg\min_\beta \sum_{i=1}^{n} (y_i - \hat{y})^2 \tag{7.44}$$

where y_i are the observed values and $\hat{y}_i$ be the predictions. Linear regression models through a straight-line fit, whereas polynomial regression models through a polynomial fit to capture all non-linear trends. Neural networks that use continuous data include activation functions specific for regression. In bioinformatics, regression analysis is used to study the relations in genetic data so that the disease likelihood could be predicted. Similarly, in climate modeling, regression is used in forecasting the temperature and precipitation based on humidity and atmospheric pressure.

7.4.5 Dimensionality Reduction

High-dimensional datasets are simplified through dimensionality reduction techniques. This is done by projecting data onto a lower-dimensional space, maintaining the essential patterns, yet reducing computational complexity. Data transformations using PCA maximize variance in the lower-dimensional space:

$$arg\ max_W W^T \sum W, \tag{7.45}$$

where W is the transformation matrix. This technique is particularly useful in bioinformatics, where complex datasets like those consisting of gene expression profiles are reduced to find the most important features that drive diseases like cancer.

7.4.6 Time Series Analysis

Time series analysis is the systematic analysis of ordered data points through time to learn about underlying trends and seasonality, as well as dependencies on temporal variables with the goal of making accurate forecasts. Two of the most celebrated methodologies are ARIMA, which combines autoregressive and moving average components to support robust time-dependent predictions, and LSTM networks that are specially tailored at modeling long-term temporal dependencies. In addition, exponential smoothing methods are often used for short-term trend forecasting. Models of time series are also important in finance applications, particularly in stock price and market index forecasting from historical trading data. Similarly, the applications of these models for smart cities lie in optimizing traffic flow management and urban planning with insights gained from history.

7.4.7 Hybrid Approaches

Hybrid approaches synergize multiple methodologies from machine learning to enhance predictive accuracy and adaptability to complex scenarios. Ensemble methods such as random forests consolidate decisions from multiple models into a stronger predictor. Hybrid strategies also often use supervised and unsupervised paradigms, which support more comprehensive and nuanced insights. Hybrid models, within the healthcare domain, have greatly enhanced the early detection of diseases by combining anomaly detection techniques to highlight rare conditions with those for the classification of conditions with desirable accuracy. The use of these different approaches, therefore, combines their complementary strengths to overcome multifaceted real-world challenges.

7.5 APPLICATIONS OF MACHINE LEARNING IN DIGITAL FORENSICS

With rapid digital data proliferation and sophistication of cybercrimes, machine learning has turned out to be an inevitable tool in the realm of digital forensics. Investigators can eventually process and analyze vast quantities of digital evidence with the ML algorithms used and can enhance the speed, accuracy, and precision of forensic inquiries accordingly. Machine learning addresses several very critical challenges related to the inherent complexity of data, the volume of information, and the need for exact event reconstruction. Therefore, it is able to offer forensic professionals a sophisticated means of uncovering hidden patterns and extracting valuable information from digital evidence.

7.5.1 Data Analysis and Evidence Extraction

Machine learning is, therefore, very fundamental in the automated analysis of large datasets, where meaningful patterns are made visible for insights that are key to forensic investigations. Through ML algorithms, one can actually extract anomalies in system logs, network activity records, and email archives. Extracting key pieces of evidence from a huge dataset is what enables investigators to speed up the process of investigations by focusing on areas of interest.

For example, through ML models, hidden relationships might be noted in system logs or user activity patterns that no one suspected to be indicative of unauthorized access and other forms of data manipulation, something that traditional approaches might not automatically help understand. Machine learning algorithms can pick anomalies with a high degree of accuracy, making the forensic analysis much more effective.

7.5.2 Malware Analysis

Malware detection and classification are among the most challenging digital forensic tasks. Machine learning provides strong techniques to detect malware, by one method, analyzing static or behavioral attributes. Under supervised learning, trained models can classify new samples based on learned characteristics as they are trained on known malware datasets. The supervised learning techniques of SVM and random forests are pretty well known in the area of malware classification.

The unsupervised learning methods, such as clustering algorithms, are also important for the identification of unknown malware variants. As these algorithms cluster files according to common behaviors, hidden or novel malware might be discovered, that otherwise would not be detected.

The following is the classification of malware often employing a combination of behavioral features, such as $x_1, x_2, \ldots, x_n$ extracted from the sample, in which each feature represents an attribute, such as an execution pattern or sequence of system calls, and a supervised model can assign a label, y 0, 1, that indicates whether the sample is benign or malicious. For example, an SVM classifier can be represented as follows:

$$y = \text{sign}\left(\sum_{i=1}^{n} w_i x_i + b\right) \tag{7.46}$$

where w_i are the weights assigned to each feature, x_i are the input features, and b is the bias term.

7.5.3 Incident Response and Intrusion Detection

Machine learning is increasingly used in IDSs to improve the detection of hostile activities in real time. The machine learning algorithm monitors and identifies network traffic and user behavior and can easily determine if there is a deviation from typical behavior, which leads to the tracking of possible security breaches. Anomaly-based detection algorithms are often used to track anomalies associated with network activity or system access for possible alerts that require further investigation.

For example, anomaly detection algorithms can very effectively identify unusual login times or unauthorized access to confidential information as a possible threat to security. Such models learn about the normal behavior of a system in action and then begin to learn what deviates from such normalcy that could represent a security event. Real-time response capabilities are part and parcel of this task because it means that forensic analysts can take action quickly, which in turn limits the potential damage a security breach could cause.

7.5.4 Timeline Reconstruction

In digital forensics, reconstructing an event sequence is an important element that can help discover causal relationships and gain a deeper understanding of sequences in cybercrime. Machine learning also assumes an important role by organizing the data systematically through various sources, such as timestamp files, user activity logs, and metadata. Basically, this process of reconstruction by automation simply improves the accuracy and efficiency in retrieving the causal order of events while providing an even stronger foundation for deriving the activities that led to the cyber incident in question.

For instance, sophisticated algorithms in analysis would detect time changes, gaps, or missing timestamps in logs and autonomously fill gaps to create a clearer timeline. This ability not only speeds up the procedures of investigation but also allows for a more accurate reconstruction of incidents, which helps minimize human error and strengthens forensic findings. Machine learning enhances the analyses of big sets of data; therefore, cybercrime investigators can have a much better understanding of the scope of cybercrime, which in turn produces more incisive conclusions. It is, therefore, necessary that the methodology for validation and assurance of digital forensic investigation be advanced.

7.5.5 Imaging and Video Forensics

Machine learning has emerged with tremendous promise in the area of imaging and video forensics, especially in the detection of modifications or manipulations of digital content. Deep learning methods, especially those involving CNNs, are widely applied to inspect the presence of anomalies within images or videos. These models are trained extensively to identify original or legitimate media and are often very good at detecting inconsistencies that, for example, may point to pixel-level modification or modified metadata.

For example, deepfake videos, which use AI to modify video content and create super-realistic yet completely fabricated depictions of events, can be very well detected using CNNs. These networks check videos for abnormalities in facial expressions, unnatural visual artifacts, or even discrepancies in motion that are not in line with the common dynamics of real human interaction. To examine these subtle discrepancies, CNNs provide a powerful tool for detecting the manipulation of visual content, thereby contributing to digital media integrity and supporting forensic investigations into digital deception.

7.5.6 Textual Evidence Analysis

NLP, a specialized subfield of machine learning, is increasingly being applied to text-based evidence, including emails and chat logs, as well as other forms of written communication. One primary application of NLP in digital forensics involves sentiment analysis, which uncovers threatening, incriminating, or otherwise suspicious content within written materials. With the analysis of the emotional tone of the text, NLP algorithms can recognize the underlying hostility, intent, or possible malicious activity and assist investigators in identifying content that could possibly denote harmful behavior or cybercrimes.

Topic Modeling: As machine learning algorithms automatically categorize and prioritize large datasets of textual information, one of the critical applications of NLP in digital forensics is topic modeling. This process allows forensic investigators to efficiently concentrate their attention on the most relevant communications and, therefore, minimize wasteful sifting through irrelevant or inconsequential information. The application of topic modeling enables investigators to systematically identify patterns, themes, or relationships that may arise within the data stream and streamline the process of investigation within forensic analysis.

7.5.7 Behavioral Profiling

Machine learning is also used to build rich behavioral profiles of users based on their interactions in the digital space. Based on these, machine learning algorithms would be able to identify out-of-the-norm patterns for login time, typing speed, file access logs, and other user activities that may denote some malicious activity. For example, an individual who usually logs in at standard business hours would raise a red flag if it's discovered that he or she is accessing sensitive files at unusual hours or performing non-routine activities.

Such profiling techniques utilize the automatic learning process of machine learning algorithms to determine a change in the behavior of users. This proves to be a robust tool in detecting insider threats, unauthorized access, and other varieties of digital malconduct. Machine learning continues to make user interactions monitored and analyzed at every point, thereby improving the speed and accuracy of behavioral anomaly detection. Consequently, it enhances the general security posture and forensic capabilities in digital environments.

7.5.8 Swarm Intelligence in Evidence Analysis

Swarm intelligence is an idea developed from computational approaches inspired by the collective behavior observed in systems like an ant colony and a bee swarm. Swarm intelligence increasingly applies to digital forensics towards improvements in various analytical tasks in optimizing processes like feature selection and evidence correlation. Through emulating the collective behavior of simple agents, the techniques of swarm intelligence, such as particle swarm optimization (PSO), enable the identification of the most relevant features within large and complex forensic datasets, thereby improving the accuracy and efficiency of machine learning models.

For example, PSO can be used to explore the optimal combination of features that most effectively differentiate between malicious and benign activities. PSO systematically finds the most informative attributes, letting the investigators better focus their attention on those important aspects of the evidence. In this way, forensic analysis speeds and precision will be enhanced. This approach not only streamlines the investigation process but also ensures that pertinent features are made available to the machine learning models, thus leading them to be useful and efficient in predictive performance with respect to cybercrime detection.

7.5.9 Predictive Analysis for Crime Prevention

Predictive analysis enables investigators to identify potential cyber threats before they manifest. It does this through historical data analysis and by recognizing recurring patterns. This makes a machine learning model capable of forecasting future crimes or system breaches. Machine learning will create proactive security measures, setting up a much better cybersecurity framework on the whole for law enforcement and forensic investigators.

For instance, through an analysis of trends in social media activity, machine learning models can identify early warning signs of cyberbullying, fraud, and other forms of online criminal behavior. Such models are created to identify the slightest patterns of anomalous behavior or escalating interactions that may presage the beginning of injurious activities. Such predictive capabilities provide great insight that one can take action on in a timely manner, thus preventing the escalation of these issues and mitigating digital crime impacts. Because this approach is anticipatory, machine learning contributes significantly to proactive threat detection and an overall enhancement of digital forensic capabilities.

7.5.10 Digital Evidence Validation

Authenticity and integrity are essential to any forensic investigation relating to digital evidence. Applying machine learning can help validate whether digital evidence has been altered or even tampered with at some stage of the investigative process.

One effective technique developed for this is cryptographic hash-based anomaly detection, involving the use of cryptographic hash functions to look for anomalies in digital data.

A cryptographic hash function produces a single, fixed-length hash value (or "digital fingerprint") of a given piece of data. Applied to digital evidence, it produces a hash value representing the exact state of the data when it was collected. Whenever there is any alteration in the data – altered intentionally or unintentionally – the produced hash value no longer matches the original one, thereby indicating possible tampering.

It ensures that the validity of the evidence is enhanced by making steady comparisons of hash values taken at successive stages of the investigation. The occurrence of a mismatch in hash values would trigger a signal suggesting the alteration of evidence, but it would be considered reliable throughout the entire process. This approach not only ensures that evidence remains uncontaminated but also ensures its admissibility in court as tampered evidence is not admissible in court. Thus, cryptographic hash-based anomaly detection serves as a crucial step in guaranteeing the authenticity of digital evidence within forensic investigations.

7.5.11 Automation of Repetitive Tasks

Machine learning, therefore can automate many of the repetitive as well as time-consuming tasks of digital forensics, giving investigators more time and resources to be deployed to the more complex and serious parts of the investigation. For instance, the use of machine learning models could automatically classify files by type or scan massive datasets for specific keywords.

For instance, machine learning algorithms can be trained to recognize and classify different types of files based on their characteristics or metadata so that investigators could quickly target pertinent files without having to manually sift through data. Likewise, through the scanning of millions of files for particular keywords or patterns, machine learning models can focus in very rapid timeframes on relevant information, such as incriminating text or evidence of malicious activity.

Automation not only accelerates forensic processes but also becomes very efficient as a great number of large datasets can be processed and examined in just about a fraction of the time required for processing through manual methods. By streamlining routine tasks, machine learning contributes to a more focused and effective investigative process as forensic professionals spend more time and expertise on higher-level analysis and decision-making.

7.5.12 Managing Big Data in Forensic Investigations

Analysis of large and complex datasets frequently results in digital forensics, making data management and processing the primary concerns. Machine learning techniques, such as dimensionality reduction and distributed computing, are crucial in addressing the challenges with respect to efficiency and scalability in forensic investigations.

One of the vital techniques used in managing large-scale data is PCA, which collapses the dimension of data while preserving the most important features. PCA identifies the principal components – key variables that capture most of the variance in the dataset – allowing the model to run more efficiently. Thus, PCA can be viewed as not only reducing the computational burden associated with big data in high dimensions but also allowing machine learning models to have an enhanced capability to conduct analysis on large amounts of data by excluding noise complexity.

Apart from dimensionality reduction, distributed computing serves as another essential way to deal with big data in digital forensics. This involves the distribution of the computing tasks across several machines, thus increasing the speed of data analysis operations. By employing parallel processing power, distributed computing allows large datasets to be processed rapidly and efficiently by forensic investigators. This allows for a high level of precision and speed when making decisions at critical points in time, such as identifying possible threats or detecting tampering in real time. It is with the combination of PCA and distributed computing that machine learning supports effortless data handling and analysis in digital forensics, which increases the speed and, thus, the accuracy of forensic investigations.

7.5.13 Inspection of Files and Formats

Machine learning can help identify and analyze different types of files, especially for cases that entail obscure or encrypted information. The power of machine learning models lies in their ability to identify hidden steganographic content within digital media, like images, by detecting patterns and inconsistencies that even human detectives might overlook. For example, CNNs and some specific algorithms can be trained to be sensitive to the tiny changes in image pixels that could indicate hidden data. Another example would be looking for proof of encrypted files by using machine learning techniques on file headers and other metadata to flag those that may be hidden or require further decryption [27].

The capabilities thus hold great worth in the domain of digital forensics; they are capable of revealing information otherwise hidden or obscured, unnoticeable if it hadn't been for such capabilities. Automating the detection of steganographic and encrypted content by machine learning significantly enhances forensic investigation efficiency and effectiveness, particularly with complex cybercrimes involving data concealment techniques.

In a nutshell, machine learning provides strong tools for dealing with great data volumes that need to be processed, complex data structures involved, and time constraints. With the implementation of machine learning in forensic workflows, significant improvements in the speed, accuracy, and reliability of analyses can be achieved for investigators. This improves the overall quality and integrity of investigations in line with forensics and allows for better detection of criminal activities while also ensuring the preservation of digital evidence.

7.6 CHALLENGES IN ANALYZING VAST DATASETS WITH ML

The high availability of datasets in multiple domains indeed brought along quite a few fantastic opportunities for applying machine learning. However, data analysis of huge dimensions is posed by quite a few challenges directly derived from complexity, sizes, and diversity. This chapter discusses these challenges along with many other dimensions, including quality, computational scalability, feature engineering, model interpretability, and size scales.

7.6.1 Data Quality and Preprocessing Challenges

Data quality is integral to the working of any machine learning model. Issues of missing values, inconsistencies, noise, and even inherent class imbalances develop when data are large. These factors add to additional difficulties in the preprocessing step and demand the development of effective data cleaning and normalization techniques.

For instance:

- **Missing Data**: The huge datasets mostly end up with missing information very often, resulting in poor models being built. Mean or median or regression-based imputation overcomes this problem but tends to bias towards specific sources of data.
- **Noisy Data**: Anomalies and errors might be masked by naturalistic patterns in the data. Analytical techniques for outlier detection include Z-scores or interquartile ranges.
- **Imbalanced Datasets**: Class imbalance skews the prediction in a classification task. Methods like oversampling, for example, SMOTE, or assigning class weights are quite important to handle such data.

7.6.2 Computational Challenges

Processing vast datasets requires significant computational resources. As dataset size increases, the following issues become more prominent:

- **Memory Constraints**: Traditional algorithms may not handle datasets that exceed system memory. Techniques like mini-batching or streaming data can alleviate this problem.
- **Computational Complexity**: Training ML models on large datasets can be computationally expensive. Optimizations such as stochastic gradient descent (SGD) or distributed computing frameworks like Apache Spark are commonly employed.
- **Latency and Real-Time Analysis**: Many applications, such as fraud detection, require real-time predictions, which are challenging with massive datasets. Low-latency systems and model optimization for inference speed are essential.

7.6.3 Feature Engineering and Dimensionality Challenges

Vast datasets often contain thousands or even millions of features, many of which may be redundant or irrelevant. This "curse of dimensionality" affects the performance of ML models by increasing computational cost and reducing model generalization.

7.6.3.1 Dimensionality Reduction Techniques

Methods like principal component analysis (PCA) or autoencoders help reduce the number of features while preserving essential information. For example, PCA projects the data onto a lower-dimensional space:

$$Z = XW, \tag{7.47}$$

where X is the data matrix and W is the matrix of eigenvectors derived from the covariance matrix of X.

7.6.3.2 Feature Selection

Techniques such as recursive feature elimination (RFE) and L1 regularization are employed to identify and retain only the most important features.

7.6.4 Model Interpretability Challenges

The complexity of machine learning models, especially those using vast datasets, can make them difficult to interpret. Black-box models like deep neural networks, while powerful, often lack transparency in their decision-making processes.

To address interpretability:

- **Explainable AI (XAI)**: Methods such as Shapley additive explanations (SHAP) or local interpretable model-agnostic explanations (LIME) help interpret model predictions by attributing importance to input features.
- **Visualization**: Heatmaps, decision trees, and partial dependence plots are used to visualize model behavior and enhance interpretability.

7.6.5 Scalability and Infrastructure Challenges

Scaling machine learning models to handle such large-scale datasets involves sturdy infrastructure and careful system design. Important considerations are given below:

- **Distributed Computing**: Systems like Hadoop and Spark allow par-parallel processing of extensive datasets and necessitate specialized knowledge for effective implementation.
- **Cloud Infrastructure**: Cloud platforms (such as AWS and Azure) provide scalable solutions to large volumes of data, but they have problems of cost management and data security.
- **Model Deployment**: In production, large models require optimization techniques, such as model quantization or pruning, to reduce source consumption.

7.6.6 Ethical and Legal Challenges

The use of vast data also raises ethical and legal questions, particularly in delicate sectors such as healthcare [28] and finance, which pose several challenges [29], namely,

- **Data Privacy**: Compliance with regulations like GDPR mandates secure. The analysis may be limited due to the management of personal data.
- **Bias and Fairness**: Bias in the training data results in discrimination. Techniques such as fairness constraints during training are essential to mitigate this issue.

7.6.7 Visualization Challenges

Visualizing insights from massive datasets is non-trivial. Effective visualization and reduction technique should preferably reduce high-dimensional data into understandable forms without oversimplifying. Examples are listed below:

- **Cluster Visualizations**: Techniques such as t-SNE or UMAP give 2D or visualization of high-dimensional clusters in 3D.
- **Scalable Dashboards**: Tools such as Tableau or a custom dashboard interaction with rich data infrastructures should enable real-time, interactive visualizations.

Analyzing vast datasets with machine learning is an intricate task requiring careful consideration of data quality, computational efficiency, feature engineering, and ethical concerns. As datasets grow in scale and complexity, innovations in algorithms, infrastructure, and explainability will continue to play a pivotal role in overcoming these challenges, ultimately driving better insights and decision-making.

7.7 CONCLUSION

Machine learning in big data really helped revolutionize the way of studying, processing, and utilizing data across a wide range of disciplines. It is through the issues, challenges, and advancements of the applications of machine learning to big data management that it can give a better understanding of the complexity of the task. Much promise has been shown, but success will depend much on overcoming huge challenges: complexity in data, computational sophistication, and ethical considerations.

For example, preprocessing a big dataset is generally full of inconsistencies, noise, and redundancy. Techniques such as normalization, quantile transformation, and outlier detection are used to maintain clean and meaningful datasets. Feature engineering is also very important to identify relevant features and perform dimensionality reduction. PCA and autoencoders can help manage high-dimensional data with minimal loss of central structures and reduced computational load.

The chapter discusses several approaches to ML on big datasets, including supervised, unsupervised, and reinforcement learning algorithms. Supervised models, such as decision trees, deep neural networks, or some forms of SVMs, were at their best in prediction and classification problems. Unsupervised techniques like k-means clustering or t-SNE are critical for pattern discovery. Reinforcement learning provides a framework for tasks related to sequential decision-making, which is particularly relevant when working with dynamic data, such as in real-time fraud detection and adaptive cybersecurity.

Scalability is a core issue when dealing with large data. Traditional models of machine learning incur huge computational overheads when it comes to large data sizes. However, the challenges posed by the aforementioned issues have become obsolete with the use of parallel processing in distributed computing frameworks such as Apache Spark and cloud services such as AWS and Google Cloud.

Terabyte-level data management is now possible without compromising factors such as cost-effectiveness and security.

Techniques such as UMAP uniform manifold approximation and projection and t-SNE encourage even higher-dimensional datasets to cause even clearer visualization of relationships and patterns. These interactive dashboards also promote decision-making by providing ML insights in formats suitable for stakeholders.

There are too many ethical issues concerning big data, which count among them data privacy, bias minimization, and compliance with regulations. Handling data involves not just processing but also encompasses a scope of building trust-based frameworks such as SHAP and LIME, which help make black-box ML models understandable and justifiable. Indeed, given the nature of future developments, existing constraints may even be compensated by new technologies such as federated learning and quantum computing. Federated learning is a method that realizes privacy-preserving collaborative model training, while quantum machine learning promises significant efficiency advantages in computation.

It can be said in simple words that the application of machine learning in the study of large datasets represents a combination of technological progress with methodological exactitude. Machine learning can tackle all problems of scale, complexity, and ethics, which allow scholars and practitioners to distill meaningful insights that facilitate progress in different contexts. Data analysis will therefore be an essential component of how to do it effectively and responsibly.

BIBLIOGRAPHY

[1] U. Srinivasulu Reddy, Aditya Vivek Thota, and A. Dharun. Machine learning techniques for stress prediction in working employees. In *2018 IEEE International Conference on Computational Intelligence and Computing Research (ICCIC)*, pages 1–4, 2018. IEEE.

[2] Moe Elbadawi, Simon Gaisford, and Abdul W. Basit. Advanced machine-learning techniques in drug discovery. *Drug Discovery Today*, 26(3):769–777, 2021.

[3] Mark Pollitt. A history of digital forensics. In Kam-Pui Chow and Sujeet Shenoi, editors, *Advances in Digital Forensics VI*, pages 3–15, Springer Berlin Heidelberg, Berlin, Heidelberg, 2010.

[4] Dominik Birk and Christoph Wegener. Technical issues of forensic investigations in cloud computing environments. In *2011 Sixth IEEE International Workshop on Systematic Approaches to Digital Forensic Engineering*, pages 1–10, 2011. IEEE.

[5] Daniel W. Woods and Sezaneh Seymour. Evidence-based cybersecurity policy? a meta-review of security control effectiveness. *Journal of Cyber Policy*, 8(3):365–383, 2023.

[6] Paul Zikopoulos, Chris Eaton, and IBM. *Understanding Big Data: Analytics for Enterprise Class Hadoop and Streaming Data.* McGraw-Hill Osborne Media, 1st edition, 2011.

[7] Sai Teja Boppiniti. Big data meets machine learning: Strategies for efficient data processing and analysis in large datasets. *International Journal of Creative Research in Computer Technology and Design*, 2(2), August 2023.

[8] Jaime G. Carbonell, Ryszard S. Michalski, and Tom M. Mitchell. 1 – An overview of machine learning. In Ryszard S. Michalski, Jaime G. Carbonell, and Tom M. Mitchell, editors, *Machine Learning*, pages 3–23. Morgan Kaufmann, San Francisco, CA, 1983.

[9] Trevor Hastie, Robert Tibshirani, and Jerome Friedman. *Overview of Supervised Learning*, pages 9–41. Springer New York, New York, NY, 2009.

[10] Horace B. Barlow. Unsupervised learning. *Neural Computation*, 1(3):295–311, 1989.

[11] Marco A. Wiering and Martijn Van Otterlo. Reinforcement learning. *Adaptation, Learning, and Optimization*, 12(3):729, 2012.
[12] Kai Arulkumaran, Marc Peter Deisenroth, Miles Brundage, and Anil Anthony Bharath. Deep reinforcement learning: A brief survey. *IEEE Signal Processing Magazine*, 34(6):26–38, 2017.
[13] Orit Hazzan and Koby Mike. Core concepts of machine learning. In *Guide to Teaching Data Science: An Interdisciplinary Approach*, pages 209–224. Springer, 2023.
[14] Qiong Wei and Roland L. Dunbrack Jr. The role of balanced training and testing data sets for binary classifiers in bioinformatics. *PloS One*, 8(7):e67863, 2013.
[15] H. Jabbar and Rafiqul Zaman Khan. Methods to avoid over-fitting and under-fitting in supervised machine learning (comparative study). *Computer Science, Communication and Instrumentation Devices*, 70(10.3850):978–981, 2015.
[16] Michael P. LaValley. Logistic regression. *Circulation*, 117(18):2395–2399, 2008.
[17] Barry De Ville. Decision trees. *Wiley Interdisciplinary Reviews: Computational Statistics*, 5(6):448–455, 2013.
[18] Shan Suthaharan and Shan Suthaharan. Support vector machine. In *Machine Learning Models and Algorithms for Big Data Classification: Thinking with Examples for Effective Learning*, pages 207–235, Springer, 2016.
[19] Gongde Guo, Hui Wang, David Bell, Yaxin Bi, and Kieran Greer. KNN model based approach in classification. In *On the Move to Meaningful Internet Systems 2003: CoopIS, DOA, and ODBASE: OTM Confederated International Conferences, CoopIS, DOA, and ODBASE 2003, Catania, Sicily, Italy, November 3–7, 2003. Proceedings*, pages 986–996, Springer, 2003.
[20] Gregor Leban, Blaž Zupan, Gaj Vidmar, and Ivan Bratko. Vizrank: Data visualization guided by machine learning. *Data Mining and Knowledge Discovery*, 13:119–136, 2006.
[21] Muhammet Fatih Ak. A comparative analysis of breast cancer detection and diagnosis using data visualization and machine learning applications. In *Healthcare*, volume 8, page 111. MDPI, 2020.
[22] Alper Sarikaya and Michael Gleicher. Scatterplots: Tasks, data, and designs. *IEEE Transactions on Visualization and Computer Graphics*, 24(1):402–412, 2017.
[23] David W. Scott. On optimal and data-based histograms. *Biometrika*, 66(3):605–610, 1979.
[24] Agnieszka Bojko. Informative or misleading? heatmaps deconstructed. In *HumanComputer Interaction. New Trends: 13th International Conference, HCI International 2009, San Diego, CA, USA, July 19–24, 2009, Proceedings, Part I 13*, pages 30–39. Springer, 2009.
[25] Jerry L. Hintze and Ray D. Nelson. Violin plots: A box plot-density trace synergism. *The American Statistician*, 52(2):181–184, 1998.
[26] S. Todd Barlow and Padraic Neville. Case study: Visualization for decision tree analysis in data mining. In *INFOVIS*, pages 149–152, 2001.
[27] S. S. Kuo, David A. Clark, and Richard Kerr. Complete package for computer automated bridge inspection process. *Transportation Research Record*, 115–115, 1994.
[28] Sara Gerke, Timo Minssen, and Glenn Cohen. Ethical and legal challenges of artificial intelligence-driven healthcare. In *Artificial Intelligence in Healthcare*, pages 295–336. Elsevier, 2020.
[29] Dinesh Kumar and Nidhi Suthar. Ethical and legal challenges of AI in marketing: An exploration of solutions. *Journal of Information, Communication and Ethics in Society*, 22(1):124–144, 2024.

8 Artificial Intelligence and Forensic Intelligence

Analyzing Social Behavior through Historical Data

Mahyuddin K. M. Nasution

8.1 INTRODUCTION

By nature, humans are designed to live in sharing [1, 2]. Humans are social, interacting with each other to be in the world and adapt to the environment. In interacting, humans do not only share modalities but also share thoughts and intelligence [3, 4]. In that way, each of them treats each other in a civilized manner and forms *individual behavior* that then develops as *social behavior* [5, 6]. Behavior that is based on personal ties to personality is flexible, with the end moving following the flow of social change [7]. A change may lead to improvement or tend to worsen. The change becomes ongoing social behavior, but humans do not realize it if they do not observe it, and it requires a way of analyzing what is called *social behavioral forensics* [8].

Humans, both as individuals and as members of society, have the potential, by sharing it, to put their intelligence into devices that help them navigate life [9]. Various human endeavors according to their interests are compositions that become sciences that fill culture throughout human history, and to a certain extent from reasoning absorbed into technology [10, 11]. In various forms and shapes, technology becomes a container where human intelligence, such as computers and machines, continues to change rapidly, maturing in the sense of behaving as *artificial intelligence* [12]. Now, artificial intelligence influences various aspects of human life. Let's call that influence producing a new study called *forensic intelligence*, making it possible to analyze social behavior following the flow of facts and realities recorded as data following a timeline [13].

8.2 BACKGROUND: BEHAVIOR, TREATMENT, AND SOURCES

Behavior is a series of actions carried out by individuals, organizations, systems, or other organisms [14]. Actions come from something that is considered; beforehand, it is well thought out and through a plan humans name it according to the intention as a word. This action is something that can be carried out, which was originally

 DOI: 10.1201/9781003512820-8

something that was forced to be done [15]. When it happens frequently, then it is something normal and becomes a habit [16]. When habits settle in memory, then naturally sometimes they appear in reflection of thought or reflex in action [17, 18]. Likewise, all downstream habits form a culture [19, 20]. Behavior is a response to treatment or stimulation, something that may be an adaptation to nature, perhaps as a pressure from rules, perhaps something that is considered good as a result of thought along with reality, or perhaps even originating from thoughts that settle as memories [21]. So, behavior appears to the surface or not during interactions between fellow humans as individuals, organizations, or governments.

The sources of behavior and treatment of individuals, organizations, governments, and societies are related documents or information spaces.

8.2.1 Individual Behavior

Humans, as individuals, build their own thoughts by having a dialogue with themselves through those thoughts [22]. That is humans who, when time gives them the opportunity, with the limits of their abilities, reach the most intimate thoughts, which form behavior. Thoughts that provide good behavior and are beneficial to their lives will have an impact on each individual, while thoughts that are not appropriate will be reduced from their memory [23]. Thoughts, in reality, are actions that are carried out with planning, or at least those actions follow the thoughts that are taking place [24]. For some individuals, those thoughts are recorded in personal notes, at least in their diaries or daily notes [25].

Individual behavior comes from the treatment for and from oneself. According to the existence of the individual, humans accept the challenge to think about everything dynamically, which gives them knowledge [26]. In this way, knowledge enriches human behavior, with that knowledge providing treatment for it and changing that behavior to something better. The behavior of recording from the beginning is the behavior of documenting all behavior and treatment [27]. The good thing is that recording behavior has existed since ancient times until now in various scratches or writings, starting from scratches on cave walls to typing in various media that exist today [28].

The thought record today is no longer entirely in the form of a physical copy. Various conveniences have provided an opportunity for individuals to record thoughts into information space with various additional information and identities that cannot be denied and are easy to trace, and the activity of recording it becomes a potential behavior: sharing information, showing off, pouring out one's heart, or other emotional attitudes. The thought in the arrangement of words according to the language used becomes the behavioral gene of each individual. Information space records not only text but also various types of information, such as tables, images, audio, and video. Each type of information is also labeled with text at least, if not with various descriptions [29].

Suppose the collection of information is in a document, symbolized by *d*. The document consists of words or $d = \{w_i | i = 1, \ldots, n\}$; directly, each document has a size of $|d| = n$ words, or by tokenizing, changing into basic words or vocabulary, d = {vj|j = 1, . . . , m}, and reducing the same word then the number of tokens is$\| d \| = m$ tokens, where m $\leq$ n [30].

Each individual will have more than one document *d*, and has the potential to build a corpus D, D = {dk|k = 1, . . . , t}, where the size of the corpus is

|D| = t, with the number of words in the corpus being

$$n(\omega_D) = \mp \sum_{k=1}^{t} nk \tag{8.1}$$

or the number of tokens or vocabularies in the corpus is

$$n(v_D) = \sum_{k=1}^{t} m_k \tag{8.2}$$

Individual behavior is recorded in personal documents, which are generally confidential [31, 32].

8.2.2 Organizational Behavior

A group of people, formally, may form an organization. An organization is a system in which its members interact with each other to achieve a goal. In other words, each member treats other members in the organization, which presents a specific behavior from the organization. Meanwhile, informal organizations such as communities are also a system but more towards loose (non-binding) interactions. Organizations generally have businesses/affairs that cause interactions between their members to follow the provisions that apply in the organization and, in general, to influence their environment. Therefore, the behavior of the organization and its members is related to the documents that become provisions or mutual agreements in the organization. Meanwhile, changes to the document are more about adaptation when there are clashes that cause the organization's achievements not to be met [33].

Likewise, individuals who are members of the organization will have records in response to the provisions that take place based on organizational documents as a stimulus. As a response, currently, documents are no longer in hardcopy form, but in softcopy, which functions as learning material for the organization to grow and develop. The strength of the organization, as behavior, depends on the fulfillment of the appropriate standards of the organization. In general, interactions within the organization in carrying out the roles of each member will form interactions of related documents and are modeled as

$$Do = \sum_{i=l}^{m} (d_k \times d_l)i \tag{8.3}$$

is an organizational corpus generated through interactions within organizations, where $d_k \times d_l$ denotes the relations between documents, potential relations about $m = \frac{1}{2} n(n-1)$, and the corpus D_o contains organizational behavior [34].

Organizational behavior is recorded in organizational documents, some of which are confidential and therefore contain the organization's strategies [35].

8.2.3 Government Behavior

Government is the organization of a country, where there is a group of people who make policies to implement legislation based on the basis of the state. The constitution, as the basis of the state, is the behavior of the state. The constitution treats everyone in the country and regulates other regulations, where the government organizes the country based on it. The government issues legislation as an interpretive policy from it. This interpretation may cause the organization of the country to be good, but it may also be the opposite. [36] Suppose the constitution document is d^u, while the statutory document d_k and other regulations as hierarchical derivatives are d_l, then the corpus structure related to government behavior is

$$D_g = \{d^u \leftarrow d_k \leftarrow d_l \leftarrow \ldots dx\} \tag{8.4}$$

the notation ←' reads "depends on," $k = 1, \ldots, m$, $l = 1, \ldots, n$, and so on, so that the picture of the regulatory dependency is like a tree rooted in the constitution [37].

Documents related to the organization of the state are evidence of the behavior of the government at that time. The recorded and implemented policies may be a picture of whether the government improves the welfare of the country or if the policy is to take unilateral advantage by the government or a group of people such as oligarchy. Disclosure of government behavior refers to documents interpreting laws, then testing with various outputs such as the rankings achieved in various determining fields, such as economic status, corruption status, and law enforcement [38]. However, the government tends to make rules to force individuals or organizations to fulfill government requests, such as rules on tax increases or imposing taxes on all activities of individuals or organizations, or vice versa to improve the welfare of citizens [39].

In general, all documents containing government behavior are transparent, while some field implementations are potentially confidential [40].

8.2.4 Social Behavior

Social elements consist of individuals, organizations, communities, and governments, or as a whole are members of society. Social behavior comes from the accumulation of behavior of social elements, where one interacts with another by giving treatment after treatment. In a democracy, treatment that causes the presence of better behavior will be generally welcomed, but, on the contrary, inappropriate treatment, such as nepotism, which is contrary to democracy, will receive ridicule and become bad behavior [41, 42].

Behavior and treatment, forming a causal relationship, in social activities and recorded in documents that can be accessed by all social members.

Sometimes, the document is in the form of news in newspapers and is read by many people, as well as news from various media such as television, radio, and publications that spread from publishers to viewers or readers in one direction. So, information flows one way, which provides treatment to all social members. However, now with the Internet and the web, information flows both ways; each individual may comment on the information they receive. Treatment is reciprocated with treatment, socially it applies as social behavior that is currently an activity in social media, which forms social behavior [43]. Status or notes recorded in social media are not only current interaction behavior, but also treatment of social elements, which invite attention from other social elements. As social behavior, there is treatment that has the potential to provide extraordinary circumstances, which describe social turmoil [44].

The interaction between social elements involving the web and social media, recording the daily activities of social elements, forms an information space in various dimensions where there are large data or information platforms that can be mined or extracted. Suppose the information space is stated as Ω, a documentation system consisting of documents or web pages, $\omega_i \in \Omega\, i = 1, \ldots, n$, as points, connected by an Internet network or link to form a document network, where edge by edge is formed from $(\omega i, \omega j) \in \Omega, j \neq i, j = 1, \ldots, n$. The information space can be accessed through search engines, such as Google, Yahoo!, Edge, or others, which summarize information about the search term *tx* entered into a search engine as a query. For example, the search term is a social behavior represented by the word "love," *tx* = "love"; then the search engine will return a set of descriptions of ± 50 words around the search term, called snippets, and hit counts are expressed as subsystems Ωx and cardinality $\|\Omega x\|$, respectively. If the search term represents any behavior, the list of snippets indirectly describes the behavior [45], while the cardinality expresses the frequency of the behavior as a description of habits [46, 47, 48].

8.3 NATURAL LAW AND BEHAVIOR

Naturally, earthquakes, floods, waves, wind, storms, landslides, heat, fires, and cold are natural behaviors that follow the laws of nature that apply to land, rivers, seas, forests, air, water, fire, or ice, or laws that are set for nature. Natural behavior becomes a treatment of its inhabitants, where humans voluntarily or are forced to respond to it with behavior or treatment [49–51]. Humans generally treat natural behavior by thinking, analyzing, and taking action, or reflexively avoiding/doing something [52].

Action or doing deeds involves verbs as naming. Analyzing has the potential to record also represented by different words about the purpose of analysis, as well as thinking. Natural behavior, in general, forms social behavior, which is recorded continuously as learning for the next generation. Humans create and carry out treatment of nature by developing thoughts about nature, and by scientists, formulate it as science, then create technology to treat nature according to human interests and manage human treatment [53].

8.3.1 Water, Words, and Deeds

The earth's surface consists of ±71% water, and the rest is land. Likewise, with humans, animals, and plants as inhabitants of the earth, each will consist of most of the water and the other part is land, following the same ratio as the earth's surface. On the other hand, words that name objects, verbs that name actions, and adjectives that express attitudes and traits affect the nature of water to be good or bad [54].

Human language, through its naming words and their meanings, is in pairs. In other words, for every word that means good, there is a word that means bad. Words with good meanings, such as the word love, affect the structure of water, becoming better in the crystal picture, but conversely, words with bad meanings, such as the word hate, cause the crystal picture of water to be damaged. The comparison of the use of good and bad words has the potential to affect the behavior of water and have an impact on life [55].

Likewise, if the act representation or represented by the word, such as committing murder or genocide, then that act not only violates human rights but also affects the behavior of water in the environment where the human rights violation occurs [56, 57].

Logically, as an analogy, if words and deeds affect water, then following the system that links all the elements of nature – water, fire, air, earth, and metal – are affected by words and deeds. In other words, individual behavior and social behavior will cumulatively affect the basic nature of nature and the behavior of the elements of nature. Following that human behavior, a person who says bad words will unconsciously or indirectly worsen the condition of the surrounding natural environment, just as a person who does bad things directly damages nature [58, 59].

8.3.2 Customs and Culture

Something that often happens or is said becomes a habit, and then becomes a culture. [60] Bad deeds, for example, such as corruption that is currently taking place (year . . . - year 2024) in Indonesia, which is carried out by most government officials, will become a habit and be considered a culture, or maybe become a culture of the Indonesian nation. Moreover, legal behavior does not prevent it and finalizes it as a mistake [61–63].

In general, a behavioral habit in a documentation system is expressed as the frequency of a word in a document, where the word frequency represents the frequency of the action, which is formulated as the number of the same word in the document, *tf* (term frequency):

$$tf(wi) = n(wi) \tag{8.5}$$

where $\omega i \in d$ is the ith word in the document, with equation (1) forming the word probability $p(wi) = n(wi)/\mid d \mid$. Or, *tf* based on vocabulary is

$$tf(vi) = n(vi) \tag{8.6}$$

where $vj \ \in d$ is the jth token in the document, with equation (2) forming the token probability $p(vi) = n(vi)/ \parallel d \parallel$. Thus, behavioral habits are expressed from smallest to largest: $tf\ (v_i) \leq tf\ (v_k) \leq \cdots \leq tf(v_m)$ [64, 65].

8.4 METHODOLOGY

Human efforts to improve the quality of life are by means of a procedure. A path, guided by knowledge and thought, to achieve goals and allow for novelty that contributes. The path in question is a methodology that characterizes scientific activities [66], which are carried out in teaching, research, and community service. Teaching to hone methodological thinking, research to analyze thinking with methodology, and service to test thinking in its methodology [67]. Thus, methodology involves facts that require any measurement, or some information extraction and data mining activities. Through rationalization, the facts raise considerations that become the way to resolve any issues about behavior [68].

8.4.1 Documents and Data

Behavior is a fact that follows human daily activities. Facts originating from individuals as social members, facts from organizational activities, and facts about government become part of facts that continuously flow socially, and as an indication of social behavior. All of that is in a collection of documents or corpus or distributed in information space. Meanwhile, by involving a measurement model against documents, behavioral data may be derived [69–71].

The criteria and strategy for data retrieval are by identifying, and involving quantification, mapping, and raising values for measuring words that represent behavior [72, 73].

1. Identification: Identifying n related documents.
2. Screening: Checking for duplicates of m documents, $\mathrm{m} \leq \mathrm{n}$, and excluding supporting documents.
3. Eligibility:k documents, $k \leq m$.
4. Inclusion as behavior, k documents.

In document collection, it is permissible to use available databases, including reputable indexing databases such as Scopus [74]. For example, corruption behavior in Indonesia, based on the Corruption Perception Index (https://www.transparency.org/en/cpi/2023), is a behavior that has received international attention [75]. After document screening and checking for eligibility, the acquisition of documents for forensics is available in the corpus. The data extracted to fulfill forensics and analyzed includes behavioral parameters, names, and year parameters [76, 77].

8.4.2 Social Behavioral Forensics

The inclusion criteria for the above data collection process state to increase forensic reliability [78].

Forensics, as a science, aims to recognize the effects of touching nature, organisms, or any event on an object or entity. So, behavioral forensics is concerned with tracing the effects of touching or treatment given to an individual or group of people. When the treatment relates to many people, behavioral forensics deals with groups of people and their interactions, which is referred to as social behavioral forensics, an accumulation of actions together or differently but in the same realm [13, 79].

When forensics involves digital data, such as tracing behavior in documents or information spaces, it is stated as social behavioral forensics, digitally showing the potential interaction between social elements, with the principle of zeros and ones to be applied in engineering the actual digital evidence for legal problems that involve technology, systems, or not [80]. The interaction not only reveals the social relationships of social actors but also states a group of people as a community that behaves specifically in society. The description of the interaction is stated as a social network [81].

Suppose $A = \{ai \mid i = 1, \ldots, n\}$ is a set of social actors. Social actors are those who have social modalities, such as education, wealth, power, and so on, that have the potential to influence others, with which relations are formed [82]:

$$A \times A \rightarrow R$$

where R is a set of relations. One way to extract social networks is to define a social network $SN = (V, E, A, R, \gamma_1, \gamma_2)$ involving information sources, based on the graph $G(V, E)$, for $V/= \phi$ is the set of vertices and E is the set of edges, where

$$\gamma_1 : A \xrightarrow{1:1} V$$

and

$$\gamma_2 : R = A \times A \rightarrow E.$$

Then expressing social behavior, perhaps through keywords or latent words, as labels of vertices and edges where the behavior is part of the social network

$$\gamma_3 : L \rightarrow B$$

so that the social network $SN = (V, E, A, R, L, B, \gamma 1, \gamma 2, \gamma 3)$, L is the latent label set, and B is the behavior set [83–86].

8.5 ROLE OF ARTIFICIAL INTELLIGENCE

Probabilistic generative models generally express a choice between a word w, a document d, and a social actor a, as a starting point for deriving the others [87]. Choose a word as a name for what humans think, which may name a social actor or name another object; indirectly, any word expresses a document written by a social actor or in the document there is a social actor as the author, and it is possible for another actor to be the object talks. Each document contains words that indicate behavior. That way there is a relationship between words and documents or vice versa, $w \rightarrow d$ or $d \rightarrow w$, briefly written as $w \leftrightarrow d$. With the same concept, $w \leftrightarrow a$, and $a \leftrightarrow d$. If each document in the corpus D exists at time T, then a timeline makes it possible to represent historical data. Figure 8.1 is a triangle shape in a time frame [88–90].

Basically, there is a relationship between three components of the model, w, d, and a, which is referred to as *triplet* $d \times a \times w$, which computationally involves many repetitions according to the number of documents, the word count of each document, and social actor searches. When each component has its own probability, namely $p(d)$ the probability of a document in the corpus, the probability of a word in each document $p(w)$, $w \in d$, and the probability of a social actor $p(a)$ in the corpus $a \in D$.

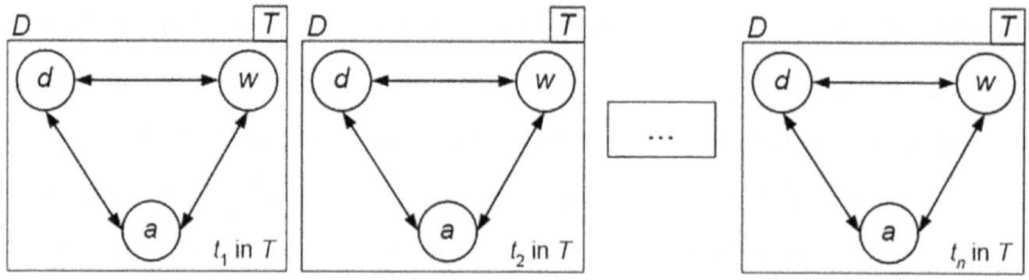

FIGURE 8.1 Generative model on timeline.

By involving Bayes' theorem, the probability of a word in a document is $p(w \mid d)$ as a conditional probability,

which is formulated as follows [91]:

$$p(w \mid d) = \frac{p(d \mid w)p(w)}{p(d)}$$

where w and d are events and $p(d) \neq 0$. The joint probability around the model of $\mathrm{d} \times a \times w$ is

$$p(d,a,w) = p(d)p(a,w \mid d)$$

involving the latent parameters of the social actor's behavior $z \in Z$ in

$$p(d,a,w) = p(d)\sum_{z \in Z} p(a,w \mid z)p(z \mid s) \tag{8.7}$$

that satisfies

1. document placement $d \in D$ with probability $p(d) \in [0, 1]$,
2. select latent class $\mathrm{z_k}$ in Z, to interpret the relationship $d \rightarrow z$, with probability $\mathrm{p(z_k|d)} \in [0, 1]$,
3. generate the word w, to interpret the relationship $z \rightarrow w$, with probability $p(w \mid z_k) \in [0, 1]$, and
4. generates the name a, to denote the relation $z \rightarrow a$, with probability $p(a|z_k) \in [0, 1]$.

Thus, the equation (8.7) provides the meaning that the latent class Z reveals that the social actor a behaves according to the word w, based on the collection of documents D. If D is a collection of documents in a year, then time $ti, i = 1, \ldots, n$, gives the trend of social behavior, where D per year gives *historical data* [92].

By duplicating human capabilities into computer-based systems, according to their interests, such as genetic algorithms, neural networks, swarm algorithms, machine learning, and others as artificial intelligence technologies. [93, 94] It is

TABLE 8.1
Number of Documents in Scopus about Three Behaviors

	All countries			Indonesia		
Year	**Corruption**	**Honest**	**Fair**	**Corruption**	**Honest**	**Fair**
2002	99	29	435			1
2003	146	29	526	2		
2004	148	39	593			
2005	211	37	708	1		1
2006	259	64	815	1		
2007	278	42	840	2		
2008	330	62	858	1		
2009	378	63	890	2		
2010	332	73	943	1		
2011	499	66	856	1		
2012	528	75	945	4		1
2013	652	69	955	4		1
2014	661	80	999	7		
2015	655	72	953	7		4
2016	823	93	955	12		2
2017	948	77	990	19		4
2018	900	94	1006	43		12
2019	946	82	1080	37		10
2020	1055	83	1222	81	1	3
2021	1038	98	1269	52		10
2022	1052	103	1406	57	1	11
2023	1073	107	1500	69	1	15

possible to describe the tendency of social behavior as a change towards better or worse from the value of $\| zi \|, i = 1, \ldots, n$ with the rate of change as the impact of behavior is

$$v_c = \frac{1}{n}\sum_{i=1}^{n} \| zi \| \tag{8.8}$$

and the acceleration of the change is

$$pc = \frac{1}{n}\sum_{i=1}^{n-1} (\| zi + i \| - \| zi \|) \tag{8.9}$$

8.6 FORENSIC INTELLIGENCE

Forensics as a science recognizes intelligence about behavior ranging from personal to social consequences. Of course, in recognizing it, exploration is carried out on matters related to behavior from the information space [95]. The behavioral cases raised, for example, are about corruption that is related to not only the personal but also social, about honesty that is related to the personal alone, or about fairness that is related to the behavior of the state or policymaking officials, Table 8.1 (based on title) [96–98]. All behaviors are also studied by the academic world that describes human interests in that behavior.

8.6.1 Corruption Personal and Social Behavior

The study of social behavior called corruption has been conducted since 1831. Since then, the scientific world has recognized corruption as human behavior until now, and it continues to be recognized as behavior that not only involves individuals but is social. The number of scientific documents about it, as evidence of the seriousness of scientists to recognize this behavior, has explored various things from a scientific perspective, and thus influenced the world that humans tend to cause damage to themselves socially [99, 100].

The study of corruption, although initially in a hesitant state to express it, but as the Western world lost its power over the Eastern world [101]. Perhaps along with that, the world is no longer confused by the idea of raping humanity, but is busy undermining a culture where certain behaviors affect human life, as shown in Figure 8.2. so as, this behavior is transformed from one culture to another; although it is not easy to transform the culture of one country to another with a different culture, personally, human life is seduced by luxury and the idea of living happily [102], until the

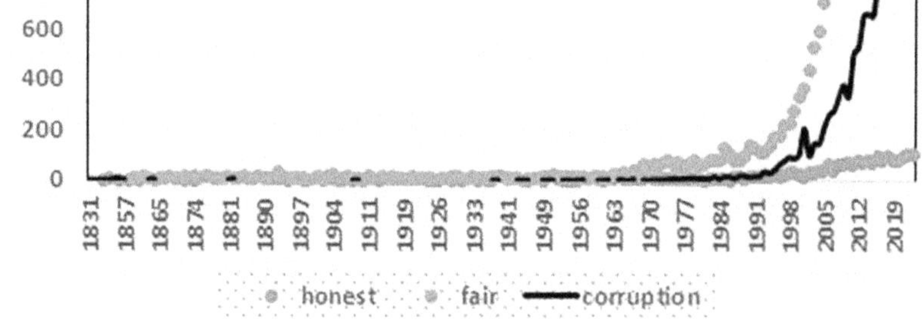

FIGURE 8.2 Document Growth about Corruption-Honest-Fair in Scopus.

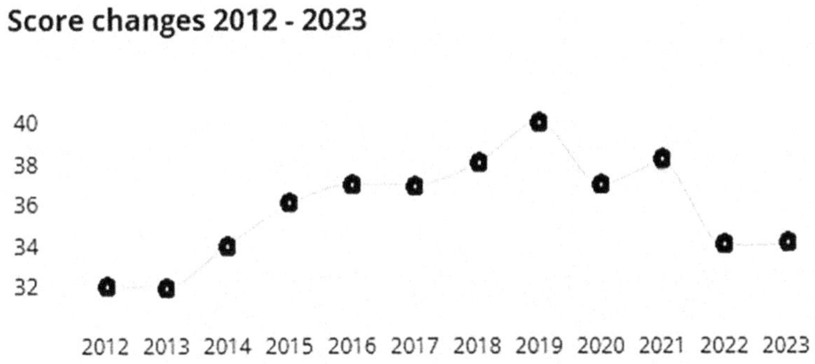

FIGURE 8.3 Corruption Perceptions Index of Indonesia [https://www.transparency.org/en/cpi/2023/index/idn].

consequences are felt in the social and state life of the Indonesian nation from 2015 to now (2024 when this book was written). Corruption has become a habit and perhaps a culture. Perhaps the Indonesian nation is the most corrupt nation in the world [103], as seen in Figure 8.3, based on the corruption perception index.

8.6.2 Community with Behavior

Behavior begins with individuals and then forms habits that apply socially. Behavior in an event may be ordinary or extraordinary; extraordinary behavior will become a public conversation and have a strong influence on social life, and the event is recorded in the information space. Exploration is carried out on documents in the information space, as an extraction that forms not only the data structure but also the structure of social behavior [104, 105].

Social behavior structure as a case. The first financial embezzlement as an extraordinary event in Indonesia [106, 107], with data extraction through Google Scholar, involved a citizen of Indonesian descent or a non-native Indonesian national named Eddy Tansil alias Tan Tjoe Hong or Tan Tju Fuan (1), who was involved in embezzlement of loan funds from the bank. Figure 8.4 shows that there were other individuals involved; followed by Hartati Murdaya (2), who was involved in bribing party cadres; then there were quite a few businessmen involved in debt embezzlement: Sudono Salim alias Liem Sioe Liong (3), Sjamsul Nursalim alias Liem Tek Siong (4), Sudwikatmono (5), Bob Hasan alias The Kian Seng (6), Usman Admadjaja (7), Ongko (8), Andrian Kiki Ariawan (9), Eko Adi Putranto (10), Sherny Kojongian (11), David Nusa Wijaya (12), and Samadikun Hartono (13); Hendra Rahardja alias Tan Tjoe Hing (14), ran away with money from Bank Indonesia; Sanyoto Tanuwidjaja (15) ran away with credit money from a government bank; Djoko Chandra alias

Tjan Kok Hui (16) in the Bank Bali scandal; Maria Pauline (17) in connection with the Bank Nasional Indonesia (BNI) burglary; Anggoro Widjojo (18) in connection with the letter from the Forestry Department and ran away; Robert Dale

	n(d)	n(d)c	dif	1	2	3	4	5	6	7	8	9	10	11	12	13	14	15	16	17	18	19	20	21	22	23	24	25	26	27	28	29
1	228	80		1	0.000	0.163	0.048	0.076	0.105	0.025	0.029	0.063	0.013	0.000	0.038	0.151	0.126	0.000	0.050	0.025	0.019	0.000	0.063	0.000	0.000	0.013	0.013	0.019	0.050	0.088	0.000	0.157
2	177	112	32	0	1	0.027	0.031	0.017	0.029	0.018	0.017	0.000	0.000	0.000	0.009	0.009	0.016	0.000	0.027	0.000	0.071	0.000	0.036	0.000	0.000	0.018	0.000	0.027	0.063	0.054	0.000	0.036
3	608	79	-33	13	3	1	0.115	0.669	0.259	0.114	0.120	0.013	0.013	0.013	0.013	0.058	0.110	0.000	0.025	0.000	0.010	0.000	0.101	0.000	0.000	0.000	0.000	0.123	0.203	0.405	0.038	0.314
4	330	227	148	11	7	26	1	0.137	0.192	0.101	0.194	0.026	0.075	0.022	0.115	0.225	0.159	0.000	0.088	0.075	0.070	0.004	0.211	0.035	0.018	0.022	0.040	0.154	0.031	0.176	0.018	0.344
5	503	118	-109	9	2	79	31	1	0.306	0.119	0.154	0.017	0.025	0.008	0.034	0.119	0.102	0.000	0.153	0.000	0.000	0.000	0.102	0.008	0.008	0.000	0.008	0.127	0.144	0.254	0.051	0.364
6	1240	343	225	36	10	89	66	105	1	0.061	0.157	0.006	0.012	0.003	0.009	0.073	0.082	0.000	0.006	0.000	0.003	0.000	0.090	0.003	0.000	0.003	0.000	0.160	0.070	0.210	0.015	0.131
7	32	12	-331	2	2	9	23	14	21	1	0.011	0.020	0.043	0.083	0.047	0.116	0.079	0.000	0.111	0.000	0.000	0.000	0.108	0.000	0.000	0.000	0.056	0.057	0.026	0.182	0.059	0.118
8	3270	175	163	5	3	21	44	27	54	2	1	0.011	0.017	0.011	0.040	0.143	0.120	0.000	0.006	0.000	0.017	0.000	0.074	0.006	0.006	0.000	0.006	0.074	0.103	0.011	0.011	0.131
9	64	50	-125	5	0	1	6	2	2	1	2	1	0.240	0.160	0.240	0.209	0.063	0.020	0.200	0.308	0.038	0.000	0.092	0.020	0.020	0.020	0.020	0.009	0.000	0.000	0.020	0.020
10	25	23	-27	1	0	1	17	3	4	1	3	12	1	0.261	0.372	0.198	0.094	0.043	0.111	0.185	0.048	0.043	0.092	0.348	0.174	0.261	0.435	0.019	0.000	0.000	0.043	0.039
11	8	8	-15	0	0	1	5	1	1	1	2	8	6	1	0.000	0.000	0.008	0.000	0.000	0.000	0.000	0.000	0.000	0.000	0.000	0.000	0.000	0.000	0.000	0.000	0.000	0.000
12	51	43	35	3	1	1	26	4	3	2	7	12	16	0	1	0.279	0.126	0.023	0.093	0.200	0.038	0.023	0.123	0.163	0.093	0.093	0.186	0.038	0.000	0.000	0.047	0.049
13	117	86	43	13	1	5	51	14	25	10	25	18	17	0	24	1	0.197	0.012	0.116	0.174	0.048	0.012	0.337	0.081	0.047	0.047	0.116	0.066	0.012	0.070	0.012	0.255
14	191	127	41	16	2	14	36	13	28	10	21	8	12	1	16	25	1	0.008	0.071	0.094	0.031	0.008	0.102	0.031	0.031	0.031	0.031	0.087	0.031	0.071	0.008	0.157
15	1	1	-126	0	0	0	0	0	0	0	0	1	1	0	1	1	1	1	0.000	0.015	0.000	0.000	0.000	0.000	0.000	0.000	0.000	0.000	0.000	0.000	0.000	0.000
16	59	27	26	4	3	2	20	18	2	3	1	10	3	0	4	10	9	0	1	0.123	0.115	0.000	0.062	0.037	0.037	0.037	0.074	0.009	0.013	0.036	0.000	0.049
17	326	65	38	2	0	0	17	0	0	0	0	20	12	0	13	15	12	1	8	1	0.038	0.015	0.077	0.123	0.062	0.092	0.169	0.028	0.013	0.000	0.000	0.010
18	114	104	39	2	8	1	16	0	1	0	3	4	5	0	4	5	4	0	12	4	1	0.010	0.029	0.029	0.019	0.048	0.038	0.019	0.000	0.010	0.000	0.000
19	1	1	-103	0	0	0	1	0	0	0	0	0	1	0	1	1	1	0	0	1	1	1	0.015	0.111	0.200	0.077	0.056	0.009	0.000	0.000	0.000	0.000
20	108	65	64	5	4	8	48	12	31	7	13	6	6	0	8	29	13	0	4	5	3	1	1	0.046	0.046	0.062	0.062	0.075	0.039	0.077	0.000	0.353
21	15	9	-56	0	0	0	8	1	1	0	1	1	8	0	7	7	4	0	1	8	3	1	3	1	0.556	0.385	0.389	0.028	0.000	0.018	0.000	0.010
22	5	5	-4	0	0	0	4	1	0	0	1	1	4	0	4	4	4	0	1	4	2	1	3	5	1	0.308	0.222	0.028	0.000	0.000	0.000	0.010
23	16	13	8	1	2	0	5	0	1	0	0	1	6	0	4	4	4	0	1	6	5	1	4	5	4	1	0.444	0.028	0.013	0.018	0.000	0.010
24	27	18	5	1	0	0	9	1	0	1	1	1	10	0	8	10	4	0	2	11	4	1	4	7	4	8	1	0.019	0.000	0.018	0.000	0.000
25	388	106	88	2	3	13	35	15	55	6	13	1	2	0	4	7	11	0	1	3	2	1	8	3	3	3	2	1	0.113	0.208	0.019	0.443
26	438	77	-29	4	7	16	7	17	24	2	3	0	0	0	0	1	4	0	1	1	0	0	3	0	0	1	0	12	1	0.143	0.026	0.176
27	309	55	-22	7	6	32	40	30	72	10	18	0	0	0	0	6	9	0	2	0	1	0	5	1	0	1	1	22	11	1	0.291	0.039
28	92	17	-38	0	0	3	4	6	5	1	2	1	1	0	2	1	1	0	0	0	0	0	0	0	0	0	0	2	2	16	1	0.020
29	360	102	85	16	4	32	78	43	45	12	23	2	4	0	5	26	20	0	5	1	0	0	36	1	1	1	0	47	18	4	2	1

FIGURE 8.4 Behavior Mapping Based on the Number of Documents.

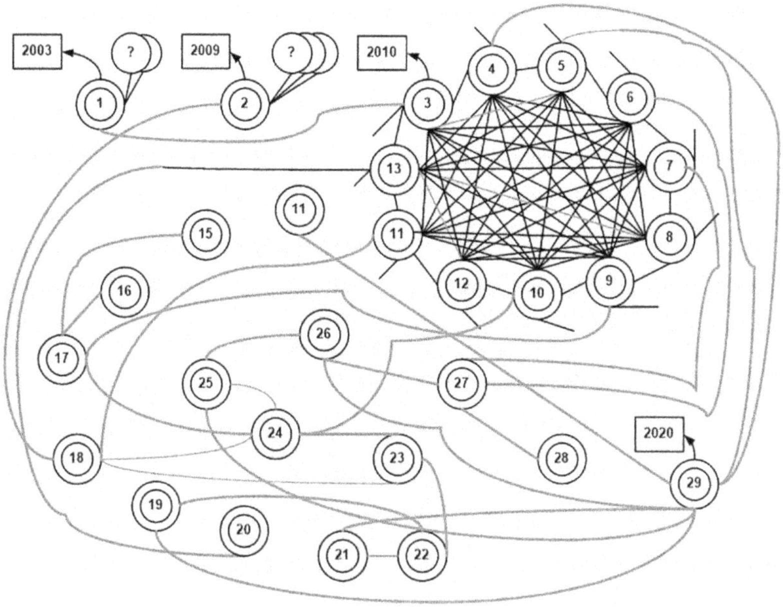

FIGURE 8.5 Social Network of the Behavior Actors Social in Timeline.

McCutchen (19), the Karaha Bodas case; Marimutu Sinivasan (20), the Bank Muamalat corruption case; Lesmana Basuki (21) and Tony Suherman (22), suspected of being involved in the Sejahtera Bank Umum corruption case; Dewi Tantular (23) and Anton Tantular (24) involved in the Bank Century case; Sukanto Tanoto (25), Unibank export money order corruption. Then they emerged as tax evaders: James Riady (26), Eka Tjipta Widjaja (27), the Salim Family (28), Sukanto Tanoto (25), and Prajogo Pangestu (29). In addition, until now, there are not a few social actors involved in the embezzlement of development funds in general [108], as shown in Figure 8.4.

Graphically, it is depicted that the behavioral point as the seed is the first social actor, plus several names that may be directly involved in the behavioral activity or not. Second, the social actor is not related to the first actor but gives behavioral influence to others, as depicted in the sequence (as a social network in the timeline). Thus, the timeline, historical data, describes the behavior that forms the influence for social and society, as shown in Figure 8.5 based on Figure 8.4, where *n*(*d*) is the number of documents in identification and *n*(*d*)*c* is the number of documents in screening and eligibility, while the number of documents on the bottom left before 1 (one) is for inclusion.

Forensic intelligence in this case is to recognize, explore, determine, and analyze possible changes from historical data extracted from the information space or a set of documents based on generative models [109–111]. The changes are based on bad or good behavior (as explained in 8.3) that have the potential to affect the continuity of life.

8.6.3 Analysis

Social behavior changes according to the message factor carried by the social actor. Then the message influences and gives an impression factor, which causes a behavior to be embedded in society, becoming social behavior. By using the equations (8.8) and (8.9), measurements of the message and its impression as the target of social behavioral forensics are generated. By considering Table 8.1 and Figure 8.2, Table 8.2 is generated, where each behavior is a vector with a weight of the number of documents, to give meaning to behavior based on the document $n(d)$ and the motion of differentiating *dif* as a way of seeing potential change in social behavior [112].

By involving document searches in occurrence and co-occurrence [113, 114], the strength relations are established involving measurement.

$$J_{overlap} = \frac{\|\Omega_x, \Omega_y\|}{\max(\|\Omega_x\|, \|\Omega_y\|} \tag{8.10}$$

To the bottom left side, we get the top right side $x_i, i = 1, \ldots, m = \frac{1}{2}(n(n-1))$ from Figure 8.4

Then involving the average measurements $\mu_x = \frac{\sum_{i=1}^{m} x_i}{m}$ and $\sigma_x = \sqrt{\frac{1}{m}\sum_{i=1}^{m}(x_i - \mu_x)^2}$, the category measurement is obtained as a follows:

$$c = \mu_x \pm \sigma_x \tag{8.11}$$

which produces categories in Table 8.3

The influence of behavior from one individual to another can be seen from the growth of relationships between individuals in the social network over a timeline.

TABLE 8.2
Message (m) and impression (im) of behavior

Range	Type	Corruption		Honest		Fair	
2002-2023	*n(d) & dif*	m	im	m	im	m	im
22 years	Documents	403	69	3	1	75	14
	Factors	18.318	3.136	0.136	0.045	3.409	0.636

TABLE 8.3
Categories

Categories	Formulation
High	$x_i \geq \mu x + \sigma_x$
Middle low	$\mu_x - \sigma x < xi < \mu x + \sigma_x xi \leq \mu x - \sigma x$

TABLE 8.4
Categories of Strength Relation in the Social Network for a Behavior

Categories	Number of x_i	Percentage
High	37	8.51%
Middle	66	15.17%
Low	332	76.32%
Total	435	100.00%

Figure 8.5 explains that the seeds of social actors who behave badly affect not only other individuals but also a group of other people, with the calculation of *n*(*d*)*c* and *dif.* The determining factor that gives an influential message to others is 74.069, and the impression is 0.759, which means that the influence of bad behavior in this case is quite large. With that impact, a little power relationship between social actors [115], the upper right side after one, and Table 8.4, what is currently happening [116], where bad behavior is socially law enforcers and policymakers in the Indonesian government as a result of bad behavior that continues to occur so that it becomes a habit and maybe even a culture [117, 118]. The answer is how that behavior has produced leaders as illegitimate children of the constitution, who are indirectly influenced by international social behavior, as the potential for the growth of documents on that behavior (Figure 8.2).

Forensic intelligence in this case is to recognize, explore, determine, and analyze *possible changes* of social behavior from historical data extracted from information space or a set of documents based on generative models. The changes are based on bad or good behavior (as explained in 8.3) that have *the potential to affect the continuity of life* of a country or a nation [119].

8.7 CONCLUSION

A behavior personally affects other individuals and then becomes social behavior if not treated. Treatment may provide something that causes behavior to become a habit and culture. Good behavior, such as honesty, has a good effect on life, and vice versa; it has an impact on the continuity of life, or not. All behavior is indirectly recorded in documents or information space and allows for tracing so that historical data supports forensic activities. Social behavior, as seen, has been revealed from appropriate documents based on applicable methodology: forensic sensibility, and reveals that the document provides messages and impressions on the continuity of life later.

BIBLIOGRAPHY

1. Tobagi, F. A., Pang, J. StarworksTM a video applications server. *1993 IEEE Compcon Spring*, 4–11, 1993.
2. Callaghan, T., Colasante, T., Muhammad, S., Corbit, J., Yavuz-Muren, M., Raffaele, C., Akter, R., Al Janaideh, R., Duan, T.-Y., Didkowsky, N., Beuze, J.-N., Homer, B. Fostering prosociality in refugee children: An intervention with Rohingya children. *Monographs of the Society for Research in Child Development*, 89(1–2), 7–109, 2024. DOI 10.1111/mono.12477

3. Ando, R., Minamizawa, K., Saraiji, M. Y. SlideFusion: Surrogacy wheelchair with implicit eyegaze modality sharing. *ACM SIGGRAPH 2020 Emerging Technologies, SIGGRAPH 2020*, 2020. DOI 10.1145/3388534.3407299
4. El-Kosairy, A., Aslan, H., Abdelbaki, N. Transforming cybersecurity: Leveraging blockchain for enhanced threat intelligence sharing. *International Journal of Safey and Security Engineering*, 14(4), 1139–1155, 2024. DOI 10.18280/ijsse.140412
5. Li, L., Zhang, Z., Liu, Z., Ye, H., He, K. Social big data oriented trust situation risk model based on individual behavior. *International Journal of Computers and Applications*, 42(2), 119–126, 2020. DOI 10.1080/1206212X.2017.1309209
6. Weber, T., Louçã, J., Gerrits, L. Dissipative structures and the relation between individual and collective aspects of social behavior. *Systems Research and Behavioral Science*, 39(2), 274–286, 2022. DOI 10.1002/sres.2777
7. Cattlin, J., Given, L. M. Information relations for social change: Exploring the information behaviour of academics undertaking impact work. *Information Research*, 29(2), 230–245, 2024. DOI 10.47989/ir292824
8. Ashraf, N., Mahmood, D., Obaidat, M. A., Ahmed, G., Akhunzada, A. Criminal behavior identification using social media forensics. *Electronics (Switzerland)*, 11(19), 2022. DOI 10.3390/electronics11193162
9. Alonso-Diaz, S. A human-like artificial intelligence for mathematics. *Mind and Society*, 23(1–2), 79–97, 2024. DOI 10.1007/s11299-024-00304-x
10. Arnaldi, S., Giardullo, P., Pelizza, A. Interesting worlds to come. Science & technology studies facing more-than-human challenges. *Tecnoscienza*, 14(2), 45–49, 2023. DOI 10.6092/issn.2038–3460/1891
11. Naya, F. New developments in communication science research in the generative AI era exploring the nature of humans and information for co-creating human-centered technologies with AI. *NTT Technical Review*, 22(11), 18–22, 2024.
12. Zinchenko, V., Mielkov, Y., Polishchuk, O., Derevinskyi, V., Trynyak, M., Iehupov, M., Salnikova, N. Human dimension of open science and the challenges of AI technologies. *E3S Web of Conferences*, 474, 2024. DOI 10.1051/e3sconf/202447402008
13. Nasution, M. K. M. Forensic intelligence: Bridging science and technology in the digital era. *AI and Emerging Technologies*, 2024. DOI 10.1201/9781003501152-4
14. Laurenti, R., Singh, J., Sinha, R., Potting, J., Frostell, B. Unintended environmental consequences of improvement actions: A qualitative analysis of systems' structure and behavior. *Systems Research and Behavioral Science*, 33(3), 381–399, 2016. DOI 10.1002/sres.2330
15. Nasution, M. K. M., Syah, R., Elveny, M. What is data science. *Data Science with Semantic Technologies: New Trends and Future Developments*, 1–25, 2023. DOI 10.1201/9781003310785–1
16. Lubis, A. R., Nasution, M. K. M., Sitompul, O. S., Zamzami, E. M. A framework of utilizing big data of social media to find out the habits of users using keyword. *ACM International Conference Proceeding Series*, 140–144, 2020. DOI 10.1145/3411174.3411195
17. Lubis, A. R., Nasution, M. K. M., Sitompul, O. S., Zamzami, E. M. Feature extraction of tweet data characteristics to determine community habits. *2022 5th International Conference on Computer and Informatics Engineering, IC2IE 2022*, 309–313, 2022. DOI 10.1109/IC2IE56416.2022.9970180
18. Kovalova, O., Shalman, T. Reading culture of teenagers in social media: A study of reading habits and perception of online literacy and educational content by teenagers. *Studies in Media and Communication*, 12(2), 26–36, 2024. DOI 10.11114/smc.v12i2.6720
19. Baumgartner, L. J. Diet and feeding habits of predatory fishes upstream and downstream of a low-level weir. *Journal of Fish Biology*, 70(3), 879894, 2007. DOI 10.1111/j.1095-8649.2007.01352.x

20. Michaelsen, P., Nyström, L., Luke, T. J., Hedesström, M. Downstream consequences of disclosing defaults: Influences on perceptions of choice architects and subsequent behavior. *Comprehensive Results in Social Psychology*, 5(1–3), 25–48, 2021. DOI 10.1080/23743603.2021.1983720
21. Gohil, M. N., Muruganantham, B., Raval, M., Bearne, L. M. Translation, cross-cultural adaptation, and validation of a Gujarati version of a theory of planned behavior questionnaire that assesses walking treatment beliefs in people with intermittent claudication. *Journal of Vascular Nursing*, 42(3), 182–190, 2024. DOI 10.1016/j.jvn.2024.05.004
22. Aisyahrani, A. I. B. A concept paper: How does spirituality affect individual's thought, emotion, behaviour, mental and physical health? *Media Publikasi Promosi Kesehatan Indonesia*, 7(10), 2390–2396, 2024. DOI 10.56338/mppki.v7i10.5969
23. Jinnai, N., Sumioka, H., Minato, T., Ishiguro, H. The impact of a humanlike communication medium on the development of intimate human relationship. *Lecture Notes in Computer Science (Including Subseries Lecture Notes in Artificial Intelligence and Lecture Notes in Bioinformatics, 10237 LNAI*, 104–114, 2017. DOI 10.1007/978-3-319-57738-8-10
24. Saloko, A., Abdulkarim, A., Sapriya, S., Kamalasari, K., Lala, A., Prayuti, Y. Digital game in developing learning model: Technology and reinforcement of students anti-corruption behavior. *Journal of Engineering Science and Technology*, 19(6), 193–200, 2024.
25. Kazis, C. A hospital diary. The night before, as though nothing had changed . . . . *Krankenpflege, Soins infirmiers*, 79(3), 58, 1986.
26. Prabaningtyas, S., Subchan, S., Fitria, I. An optimal control problem of SECR corruption dynamic model. *Journal of Advanced Research in Applied Science and Engineering Technology*, 45(2), 128–141, 2025. DOI 10.37934/araset.45.2.128141
27. Crane, C., Hotton, M., Shelemy, L., Knowles-Bevis, R. The association between individual differences in motivational readiness at entry to treatment and treatment attendance and outcome in cognitive behaviour therapy: A systematic review. *Cognitive Therapy and Research*, 48(6), 1066–1089, 2024. DOI 10.1007/s10608-024-10504-x
28. Saksa, J. R., Cohen, S. J., Srihari, V. H., Woods, S. W. Cognitive behavior therapy for early psychosis: A comprehensive review of individual vs. group treatment studies. *International Journal of Group Psychotherapy*, 59(3), 357–383, 2009. DOI 10.1521/ijgp.2009.59.3.357
29. Orsso, C. E., Gormaz, T., Valentine, S., Trottier, C. F., Matias de Sousa, I., Ferguson-Pell, M., Johnson, S. T., Kirkham, A. A., Klein, D., Maeda, N, Mota, J. F., Neil-Sztramko, S. E., Quintanilha, M., Salami, B. O., Prado, C. M. Digital intervention for behaviour change and chronic disease prevention (DIRECTION): Study protocol for a randomized controlled trial of a web-based platform integrating nutrition, physical activity, and mindfulness for individuals with obesity. *Methods*, 231, 4554, 2024. DOI 10.1016/j.ymeth.2024.09.009
30. Nasution, M. K. M., Hardi, M., Sitepu, R., Sinulingga, E. A method to extract the forensic about negative issues from Web. *IOP Conference Series: Materials Science and Engineering*, 180(1), 2017. DOI 10.1088/1757-899X/180/1/012241
31. Nasution, M. K. M., Noah, S. A. Extraction of academic social network from online database. *2011 International Conference on Semantic Technology and Information Retrieval, STAIR 2011*, 64–69, 2011. DOI 10.1109/STAIR.2011.5995766
32. Nasution, M. K. M., Herawati, E., Elveny, M. Semantic information retrieval models. *Data Science with Semantic Technologies: Deployment and Exploration*, 49–72, 2023. DOI 10.1201/9781003310792-4
33. Juliana, A. P., Lemy, D. M., Pramono, R., Situmorang, J. M. H., Djakasaputra, A. What drives sustainable organizational citizenship behavior. *International Journal of Sustainable Development and Planning*, 17(5), 1403–1410, 2022. DOI 10.18280/ijsdp.170503
34. Malik, S. Data-driven decision-making: Leveraging the IoT for RealTime sustainability in organizational behavior. *Sustainability (Switzerland)*, 16(15), 2024. DOI 10.3390/su16156302

35. Olguin, D. O., Waber, B. N., Kim, T., Mohan, A., Ara, K. Sensible organizations: Technology and methodology for automatically measuring organizational behavior. *IEEE Transactions on Systems, Man, and Cybernetics, Part B: Cybernetics*, 39(1), 43–55, 2009. DOI 10.1109/TSMCB.2008.2006638
36. Bynum, S. A., Malo, T. L., Lee, J.-H., Guiliano, A. R., Vadaparampil, S. T. HPV vaccine information-seeking behaviors among us physicians: Government, media, or colleagues? *Vaccine*, 29(32), 5090–5093, 2011. DOI 10.1016/j.vaccine.2011.04.134
37. Park, M., Oh, K., Kim, H., Jun, J., Giap, T.-T.-T. Information-seeking and risk perception to explain infection-prevention behaviors: Conditional analysis on trust in media and government as moderator. *Patient Preference and Adherence*, 18, 2135–2145, 2024. DOI 10.2147/PPA.S484216
38. Tobing, M. H., Sirait, N. N., Siregar, M. Contiguity of law violations in the context of business competition and corruption in bid rigging cases. *Journal of Ecohumanism*, 3(4), 476–483, 2024. DOI 10.62754/joe.v3i8.4748
39. Wang, G., Chen, Q., Xie, S. Determining the types of internet plus government platform usage behaviour in selected cities of China. *Electronic Library*, 39(5), 201–221, 782–799. DOI 10.1108/EL-06-2021-0117
40. Xie, Q., Wu. H., Zhang, R. Using online negative emotions to predict risk-coping behaviors in the relocation of Beijing municipal government. *Scientific Reports*, 14(1), 2024. DOI 10.1038/s41598-024-79846-6
41. Li, N., Guo, X., Chen, G., Luo, N. Reading behavior on intraorganizational blogging systems: A group-level analysis through the lens of social capital theory. *Information and Management*, 52(7), 870–881, 2015. DOI 10.1016/j.im.2015.03.004
42. Cano Murillo, D. E., Kang, J., Yoon, S. Factors influencing prosocial consumer behavior through non-profit organizations. *Internet Research*, 26(1), 626–643, 2016. DOI 10.1108/IntR-10-2014-0253
43. Machmud, M., Fatimah, J. M., Sultan, M. I., Farid, M. Social media as communication tools for anti-corruption campaign in Indonesia. *International Journal of Data and Network Science*, 8(1), 357–368, 2024. DOI 10.5267/j.ijdns.2023.9.018
44. Ahmad, S. Political behavior in virtual environment: Role of social media intensity, internet connectivity, and political affiliation in online political persuasion among university students. *Journal of Human Behavior in the Social Environment*, 30(4), 457–473, 2020. DOI 10.1080/10911359.2019.1698485
45. Nasution, M. K. M., Noah, S. A. Superficial method for extracting social network for academic using web snippets. *Lecture Notes in Computer Science (including subseries Lecture Notes in Artificial Intelligence and Lecture Notes in Bioinformatics), 6401 LNAI*, 483–490, 2010. DOI 10.1007/978-3-642-16248-0 68
46. Nasution, M. K. M., Syah, R., Elveny, M. Studies on behaviour of information to extract the meaning behind the behaviour. *Journal of Physics: Conference Series*, 801(1), 2017. DOI 10.1088/1742-6596/801/1/012022
47. Nasution, M. K. M., Syah, R., Elfida, M. Information retrieval based on the extracted social network. *Advances in Intelligent Systems and Computing*, 662, 220–226, 2018. DOI 10.1007/978-3-319-67621-0_20
48. Nasution, M. K. M., Hidayat, R., Syah, R. Computer science. *International Journal on Advanced Science Engineering Information Technology*, 12(3), 1142–1159, 2022. DOI 10.18517/ijaseit.12.3.14832
49. Hong, X., Manafian, J., Ilhan, O. A., Alkireet, A. I. A., Nasution, M. K. M. Multiple soliton solutions of the generalized Hirota-SatsumaIto equation arising in shallow water wave. *Journal of Geometry and Physics*, 170, 2021. DOI 10.1016/j.geomphys.2021.104338
50. Syah, R., Alizadeh, S. M., Nurgalieva, K. S.Guerrero, J. W. G., Nasution, M. K. M., Davarpanah, A., Ramdan, D., Metwally, A. S. M. A laboratory approach to measure enhanced gas recovery from a tight gas reservoir during supercritical carbon dioxide injection. *Sustainability (Switzerland)*, 13(21), 2021. DOI 10.3390/su132111606

51. Syah, R., Faghri, S., Nasution, M. K. M., Davarpanah, A., Jasczur, M. Modeling and optimization of wind turbines in wind farms for solving multi-objective reactive power dispatch using a new hybrid scheme. *Energies*, 14(18), 2021. DOI 10.3390/en14185919
52. Lubis, I., Nasution, M. K. M. Probability model for designing environment condition. *Journal of Physics: Conference Series*, 801(1), 2017. DOI 10.1088/1742-6596/801/1/012026
53. Nasution, M. K. M., Sitompul, D., Harahap, M. Modeling reliability measurement of interface on information system: Towards the forensic of rules. *IOP Conference Series: Materials Science and Engineering*, 308(1), 2017. DOI 10.1088/1757-899X/308/1/012042
54. Nasution, M. K. M. World on data perspective. *World*, 3(3), 736–752, 2022. https://doi.org/10.3390/world3030041
55. Emoto, M. Healing with water. *Journal of Alternative and Complementary Medicine*, 10, 19–21, 2004. DOI 10.1089/1075553043228489
56. Amin, M. Towards self-healing energy infrastructure systems. *IEEE Computer Applications in Power*, 14(1), 20–28, 2001. DOI 10.1109/67.893351
57. Jabr, S., Berger, E. Palestine meeting Gaza's mental health crisis. *The Lancet Psychiatry*, 11(1), 12, 2024. DOI 10.1016/S2215-0366(23)00398X
58. O'Cass, A., Fenech, T. Web retailing adoption: Exploring the nature of internet users Web retailing behaviour. *Journal of Retailing and Consumer*, 10(2), 81–94, 2003. DOI 10.1016/S0969-6989(02)00004-8
59. Forstmann, M., Sagioglou, C. Lifetime experience with (classic) psychedelics predicts pro-environmental behavior through an increase in nature relatedness. *Journal of Psychopharmacology*, 31(8), 975–988, 2017. DOI 10.1177/0269881117714049
60. Yuan, H. Design of folk custom traditional culture resources optimization system based on cloud platform. *Proceedings of SPIE The International Society for Optical Engineering*, 12474, 2022. DOI 10.1117/12.2653730
61. Purnamasari, P., Amran, N. A, Nu'man, A. H., Frendika, R., Nor, M. N. M., Ismail, M. S. Penta-Helix model of E-government in combating corruption in Indonesia and Malaysia: Religiosity as a moderating role. *F1000Research*, 11, 932, 2022. DOI 10.12688/f1000research.121746.1
62. Hidayat, S., Amiruddin, A., Parman, L. Reformulation of certain circumstances indicators as a prerequisite for the imposition of death penalty in corruption crime. *Jurnal IUS Kajian Hukum dan Keadilan*, 12(1), 43–54, 2024. DOI 10.29303/ius.v12i1.1350
63. Rahman, A. M., Husnul, M. Failure of criminal law in recovering state losses due to criminal acts of corruption. *Petita: Jurnal Kajian Ilmu Hukum dan Syariah*, 9(1), 305–323, 2024. DOI 10.22373/petita.v9i1.244 Or see URL: https://petita.ar-raniry.ac.id/index.php/petita/article/view/244
64. Lubis, A. R., Nasution, M. K. M., Sitompul, O. S., Zamzami, E. M. The effect of the TF-IDF algorithm in times series in forecasting word on social media. *Indonesian Journal of Electrical Engineering and Computer Science*, 22(2), 368–376, 2020. DOI 10.11591/ijeecs.v22.i2.pp368-376
65. Lubis, A. R., Nasution, M. K. M., Sitompul, O. S., Zamzami, E. M. The effect of the TF-IDF algorithm in times series in forecasting word on social media. *Indonesian Journal of Electrical Engineering and Computer Science*, 22(2), 976–984, 2021. DOI 10.11591/ijeecs.v22.i2.pp976-984
66. Nasution, M. K. M. Methodology. *Journal of Physics: Conference Series*, 1566(1), 2020. DOI 10.1088/1742-6596/1566/1/012031
67. Nasution, M. K. M., Elveny, M., Syah, R. Research methodology for computer science. *AIP Conference Proceedings*, 2714, 2023. DOI 10.1063/5.0134511
68. Nasution, M. K. M., Elveny, M., Syah, R., Noah, S. A. Behavior of the resources in the growth of social network. *Proceedings 5th International Conference on Electrical Engineering and Informatics: Bridging the Knowledge between Academic, Industry, and Community, ICEEI 2015*, 496–499, 2015. DOI 10.1109/ICEEI.2015.7352551

69. Zhu, P., Lu, X., Zhou, X. Web document pre-sending based on user behavior patterns. *Ruan Jian Xue Bao/Journal of Software*, 10(11), 11421147, 1999.
70. Freeman, E., Passos, K., Wuthrich, C., Stringfellow, C. Using sequence diagrams to document web elements' unknown behavior. *2004 IEEE Wireless Communications and Networking Conference, WCNC 2004*, 4, 2587–2592, 2004.
71. Ji, J., Kunita, D., Zhao, Q. A human behavior inspired awareness system for document analysis. *2010 2nd International Symposium on Aware Computing, ISAC 2010 Symposium Guide*, 185–190, 2010. DOI 10.1109/ISAC.2010.5670477
72. Nasution, M. K. M., Noah, S. A. Information retrieval model: A social network extraction perspective. *Proceedings 2012 International Conference on Information Retrieval and Knowledge Management, CAMP'12*, 322–326, 2012. DOI 10.1109/InfRKM.2012.6204999
73. Elveny, M., Syah, R., Elfida, M., Nasution, M. K. M. Information retrieval on social network: An adaptive proof. *IOP Conference Series: Materials Science and Engineering*, 300(1), 2018. DOI 10.1088/1757899X/300/1/012055
74. Gusenbauer, M. Beyond Google Scholar, Scopus, and Web of Science: An evaluation of the backward and forward citation coverage of 59 databases' citation indices. *Research Synthesis Methods*, 15(5), 802–817, 2024. DOI 10.1002/jrsm.1729
75. Sari, A. E-government attempts in small Island developing states: The rate of corruption with virtualization. *Science and Engineering Ethics*, 23(6), 1673–1688, 2017. DOI 10.1007/s11948-016-9848-0
76. Nasution, M. K. M., Sitompul, O. S. Enhancing extraction method for aggregating strength relation between social actors. *Advances in Intelligent Systems and Computing*, 573, 312–321, 2017. DOI 10.1007/978-3-319-57261-1_31
77. Lubis, A. R., Nasution, M. K. M., Sitompul, O. S., Zamzawi, E. M. The feature extraction for classifying words on social media with the Naïve Bayes algorithm. *IAES International Journal of Artificial Intelligence*, 11(3), 1041–1048, 2022. DOI 10.11591/ijai.v11.i3.pp1041-1048
78. Tanga, C., López-Lázaro, S., Sato-Álvarez, C., Viciano, J. Reliability of predictive models based on the mesiodistal and buccolingual diameters of permanent canine teeth for sex estimation in forensic contexts: A systematic review and meta-analysis. *Forensic Science International*, 361, 2024. DOI 10.1016/j.forsciint.2024.112143
79. Rogers, R., Tazi, K. Y., Dragin, E. Y. Forensic assessment instruments: Their reliability and applicability to criminal forensic issues. *Behavioral Sciences and the Law*, 41(5), 415–431, 2023. DOI 10.1002/bsl.2613.
80. Burton, B., Dragin, E. Y. Acquisition and interpretation of digital information in forensic assessments. *Harvard Review of Psychiatry*, 29(3), 234–239, 2021. DOI 10.1097/HRP.0000000000000291
81. Al-Khateeb, S., Hussain, M., Agarwal, N. Analyzing deviant sociotechnical behaviors using social network analysis and cyber forensics based methodologies. *Big Data Analytics in Cybersecurity*, 263–280, 2017. DOI 10.1201/9781315154374-12
82. Nasution, M. K. M. Social network mining (SNM): A definition of relation between the resources and SNA. *International Journal on Advanced Science, Engineering and Information Technology*, 6(6), 975–981, 2016. DOI 10.18517/ijaseit.6.6.1390
83. Nasution, M. K. M., Noah, S. A. Social network extraction based on Web. A comparison of superficial methods. *Procedia Computer Science*, 124, 86–92, 2017. DOI 10.1016/j.procs.2017.12.133
84. Nsution, M. K. M. Social network extraction based on Web: 1. Related superficial methods. *IOP Conference Series: Materials Science and Engineering*, 300(1), 2018. DOI 10.1088/1757-899X/300/1/012056
85. Nasution, M. K. M. Social network extraction based on web: 2. Strategies in superficial methods. *Journal of Physics: Conference Series*, 1116(2), 2018. DOI 10.1088/1742-6596/1116/2/022029

86. Nasution, M. K. M., Sitompul, O. S., Noah, S. A. Social network extraction based on Web: 3. The integrated superficial method. *Journal of Physics: Conference Series*, 978(1), 2018. DOI 10.1088/17426596/978/1/012033
87. Lau, R. Y. K., Xia, Y., Ye, Y. A probabilistic generative model for mining cybercriminal networks from online social media. *IEEE Computational Intelligence Magazine*, 9(1), 31–43, 2014. DOI 10.1109/MCI.2013.2291689
88. Song, J., Lu, X., Hong, J., Wang, F. External information enhancing topic model based on graph neural network. *Expert Systems with Applications*, 263, 2025. DOI 10.1016/j.eswa.2024.125709
89. Zeng, H., Sun, J.-M., Li, C.-S., Li, Z., Wei. T. Foundation models for topic modeling: A case study. *Frontiers of Computer Science*, 19(2), 2025. DOI 10.1007/s11704-024-40069-7
90. Li, N., Lv, T., Wang, X., Meng, X., Xu, J., Guo, Y. Research progress and hot topics of distributed photovoltaic: Bibliometric analysis and Latent Dirichlet allocation model. *Energy and Buildings*, 327, 2025. DOI 10.1016/j.enbuild.2024.115056
91. Koocherneshkian, P., Bouguila, N. Integration of neural embeddings and probabilistic models in topic modeling. *Applied Artificial Intelligence*, 38(1), 2024. DOI 10.1080/08839514.2024.2403904
92. Murakami, A., Hunstan, S., Thompson, P. Contextualising topic models in corpus linguistics: Opportunities and challenges. *Discourse Studies*, 2024. DOI 10.1177/14614456241293896
93. Alhashmi, S. M., Hashem, I. A. T., Al-Qudah, I. Artificial intelligence applications in healthcare: A bibliometric and topic model based analysis. *Intelligent Systems with Applications*, 21, 2024. DOI 10.1016/j.iswa.2023.200299
94. Masrom, S., Septiyanti, R., Ahmad, A., Rahman, R. A., Sulaiman, N. Analysis of machine learning in classifying bank profitability with corruption factor. *Journal of Advanced Research in Applied Sciences and Engineering Technology*, 40(2), 13–21, 2024. DOI 10.37934/araset.40.2.1321
95. Rizal, R., Selamat, S. R., Mas'ud, M. Z., Widiyasono, N. Enhanced readiness forensic framework for the complexity of internet of things (IoT) investigation based on artificial intelligence. *Journal of Advanced Research in Applied Sciences and Engineering Technology*, 50(1), 121135, 2025. DOI 10.37934/araset.50.1.121135
96. Beerepoot, I., Šinik, T., Reijer, H. A. Capturing and analysing employee behaviour: An honest day's work record. *Data and Knowledge Engineering*, 154, 2024. DOI 10.1016/j.datak.2024.102350
97. Mok, C., Kim, S. B. Corruption-based anomaly detection and interpretation in tabular data. *Pattern Recognition*, 159, 2024. DOI 10.1016/j.patcog.2024.111149
98. Wang, Y., Tan, G., Si, H. Decision modeling for automated driving in dilemmas based on bidirectional value alignment of moral theory values and fair human moral values. *Transportation Research Part F: Traffic Psychology and Behaviour*, 108, 14–27, 2025. DOI 10.1016/j.trf.2024.11.001
99. Luo, Y. Political behavior, social responsibility, and perceived corruption: A structuration perspective. *Journal of International Business Studies*, 37(6), 747–766, 2006. DOI 10.1057/palgrave.jibs.8400224
100. Brown, R. C. H. Social values and the corruption argument against financial incentives for healthy behaviour. *Journal of Medical Ethics*, 43(3), 140–144, 2017. DOI 10.1136/medethics-2016-103372
101. Vilaça, L. A social movement model for judicial behavior: Evidence from Brazil's anti-corruption movements. *Social Forces*, 103(1), 223–244, 2024. DOI 10.1093/sf/soae065
102. Prabowo, H. Y. Where is the life we have lost in living (beyond means)? An exploratory inquiry into the deceptive world of corruption and consumerism. *Journal of Financial Crime*, 31(6), 1540–1560, 2024. DOI 10.1108/JFC-10-2023-0273

103. Hufron, H., Fikri, S. The urgency of regulating forfeiture of assets gained from corruption in indonesia. *Legality: Jurnal Ilmiah Hukum*, 32(2), 292310, 2024. DOI 10.22219/ljih.v32i2.35243
104. Nasution, M. K. M. Mining of the extracted social network. *Journal of Theoretical and Applied Information Technology*, 99(13), 3219–3230, 2021.
105. Nasution, M. K. M. A study about complexity of social network. *Lecture Notes in Networks and Systems, 1119 LNNS*, 158–170, 2024. DOI 10.1007/978-3-031-70300-3_11
106. Nasution, M. K. M. Social network mining: A discussion. *Journal of Physics: Conference Series*, 1235(1), 2018. DOI 10.1088/17426596/1235/1/012111
107. Utami, E. R., Barokah, Z. The determinants of corporate anticorruption disclosures: evidence from construction companies in the Asia-Pacific. *Corporate Governance (Bingley)*, 24(6), 1414–1441, 2024. DOI 10.1108/CG-04-2023-0152
108. Hansen, S. Involvenemtn of government officials in the construction sector corruption. *Journal of Legal Affairs and Dispute Resolution in Engineering and Construction*, 16(2), 2024. DOI 10.1061/JLADAH.LADR1014
109. Nasution, M. K. M., Hardi, M., Sitepu, R. Using social networks to assess forensic of negative issues. *Proceedings of 2016 4th International Conference on Cyber and IT Service Management, CITSM 2016*, 2016. DOI 10.1109/CITSM.2016.7577513
110. Nasution, M. K. M., Hardi, M., Sitepu, R., Sinulingga, E. A method to extract the forensic about negative issues from web. *IOP Conference Series: Materials Science and Engineering*, 180(1), 2017. DOI 10.1088/1757-899X/180/1/012241
111. Nasution, M. K. M. Forensic in information technology: A redefinition. *Journal of Physics: Conference Series*, 1235(1), 2019. DOI 10.1088/1742-6596/1235/1/012106
112. van Dongen, J. D. M., Drislane, L. E., Nijman, H., Soe-Agnie, S. E., van Marie, H. J. C. Further evidence for reliability and validity of the Triarchic psychopathy measure in a forensic sample and a community sample. *Journal of Psychopathology and Behavioral Assessment*, 39(1), 58–66, 2017. DOI 10.1007/s10862-016-9567-5
113. Nasution, M. K. M. Singleton: A role of the search engine to reveal the existence of something in information space. *IOP Conference Series: Materials Science and Engineering*, 420(1), 2018. DOI 10.1088/1757899X/420/1/012137
114. Nasution, M. K. M. Doubleton: A role of the search engine to reveal the existence of relation in information space. *IOP Conference Series: Materials Science and Engineering*, 420(1), 2018. DOI 10.1088/1757899X/420/1/012138
115. Rifai, R., Kamaluddin, K., Hidayat, R. Exploring the roots and solutions of maladministration, power abuse, or corruption in contemporary Indonesian villages. *Pertanika Journal of Social Sciences and Humanities*, 32(3), 1115–1140, 2024. DOI 10.47836/pjssh.32.3.15
116. Prastica, R. M. S., Prasetya, N. D., Setyaasri, Y. P. What are the triggering factors of corruption in construction industry in Indonesia? *AIP Conference Proceedings*, 3145(1), 2022. DOI 10.1063/5.0214112
117. Yunan, Z. Y., Alharthi, M., Jeris, S. S. The risk of political instability and the performance of Islamic banks: Does corruption matter? *Journal of Financial Crime*, 31(6), 1384–1406, 2024. DOI 10.1108/JFC-09-2023-0229
118. Marlina, A., Haq, I., Karauwan, D. E. S., Waeno, M. Unveiling the ripple effects of corruption in education sector: A solution from Islamic perspective. *Petita: Jurnal Kajian Ilmu Hukum dan Syariah*, 9(3), 762776, 2024. DOI 10.22373/PETITA.V9I2.283
119. Lestari, W. R., Sunarmi, Wahyudi, H. Relationship corruption and economic growth ASEAN countries. *Journal of Ecohumanism*, 3(7), 29262939, 2024. DOI 10.62754/joe.v3i7.4427

9 Identifying IOCs from Compromised Android Devices

Prince Bhuva and Archana Patel

9.1 INTRODUCTION

Malware is a threat that affects computer systems, applications, and networks. They are programs that disrupt the smooth running of a computer system, application, or network. There are various types of malware ranging from rootkits, viruses, Trojans, backdoors, and more [1]. The increase in the use of Android smartphones attracts malicious threat actors. Hence, the security of Android devices has also become a critical issue. Detection and prevention of malware in Android devices is of great concern as they rapidly add to the ecosystem of the tech world and thus create a broader attack surface for threat actors.

With the above concern in mind, it is proposed to develop a technique for the identification of IOCs from compromised Android devices, which can be used, by security researchers, malware analysts, and law enforcement agencies to observe the characteristics and behavior of devices after being affected by malware.

Nowadays malware is the center of attraction for threat actors. The increase in the use of Android smartphones attracts malicious threat actors. Hence, the security of Android devices has also become a critical issue. Detection and prevention of malware in Android devices is of great concern. Today, most malware is a combination of traditional malicious code; it often includes Trojans, some parts of a worm, and sometimes a virus.

9.1.1 MALWARE

In simple words, malware is a name for malicious software or code that is used to interrupt the functionality of programs, infrastructure, or networks. It delivers payloads to the respective system using different ways, such as infecting APK files, PDF files, etc. Malware steals information from the infected system without the knowledge of the user. There are many types of malware; we can divide them into different types based on their behavior [2].

- Virus: A virus is attached to the original code of a file and makes it malicious; after that, it triggers when the user uses that specific infected function or process. A virus can replicate itself by changing other programs and inserting its own code.

DOI: 10.1201/9781003512820-9

- Worms: A worm is a malware program that replicates itself and can spread to other systems. Generally, it uses the network to spread itself. It will use the infected machine as a host machine to scan and infect other systems.
- Trojan: A Trojan is malware that seems genuine but is capable of taking control of the whole system. Using that characteristic, it can trick the user. It requires the user to click on it, after that loads, and runs that code on the device. It will create backdoors for malicious code, which will give them easy access to enter the system.
- Malware Bots: A malware bot is a type of program designed to infiltrate user accounts and scan the Internet for personal information, which can then be used to cause harm. It affects the user's computer using download formats through social media or email messages that notify users to click on a link.
- Spyware: Spyware is malware that monitors or tracks the infected user's activity to collect information about the user. It behaves by gathering information from the infected user and sending it to a third party.
- Ransomware: Ransomware is malware that can encrypt user data in such a way that it cannot be accessed by them until they pay the ransom. In another situation, it can lock the system as well. In that case, a person with the right skills can easily reverse it. Ransomware is generally attached to an email attachment and is typically carried out by a Trojan.

9.1.2 Overview of Android

The Android operating system uses Linux as its base. Android developers develop applications using Java libraries, which are actually designed by Google. The Linux kernel provides services like management at runtime and core service handling. The libraries included are same as those in Linux like Libc and SSL. The libraries included are the same as those in Linux, such as Libc and SSL. Android Runtime includes Dalvik, a virtual machine, along with some core libraries. On the other side, the App Framework provides classes that can create apps, abstractions for hardware access, and manages user interface resources. All applications in Android are made up on the Application Layer. By understanding the Android architecture, we geta better idea of its components and other parts.

9.1.3 Malware Analysis Methods

Malware analysis is a method of learning the functionality and behavior of malware samples. Through this process, we can identify the types, behavior, and methodologies of the malware. There are two types: static malware analysis and dynamic malware analysis [3].

- Static Analysis: Static malware analysis is a technique in which we analyze the malware sample without running the sample. It has tools and techniques by which we can say that a given file is malicious or not. It is further classified into two categories: basic static analysis and advanced static analysis. In basic static analysis, we analyze our sample using various tools like

VirusTotal, where it is checked by anti-virus solutions. The strings of the code are checked using tools like BinText and PEiD to determine whether they are malicious. In advanced static analysis, we can perform reverse engineering of the malware sample's source code to determine whether it is malicious.
- Dynamic Analysis: Dynamic malware analysis is a method in which a malware sample is executed to observe its behavior and the effects it causes, using system internal tools. Through this technique, we can get an exact idea of how actually the malware behaves and what type of damage it causes to the system.

The aim of this research is to develop a technique for the identification of IOCs from compromised Android devices for law enforcement agencies, SoC teams, and security researchers for easy identification of malware-infected devices. To achieve the aim of this chapter, it is of great importance to ensure the following objectives are taken into consideration:

- Identification of indicators of compromise from Android devices
- Analyze various types of APKs and perform the static and dynamic analysis on them

This chapter includes the development of a technique for the identification of IOCs from compromised Android devices using malware analysis techniques. It also includes the configuration method of the Sandbox and comparative analysis of devices.

9.2 LITERATURE REVIEW

The following work has been carried out by researchers:

- Khaled Bakour et al. [4]: The Android Malware Detection System between Hope and Reality: They used a schematic review model to illustrate the process of malware detection. Challenges with the static and dynamic analysis include tackling obfuscation technique and time complexity.
- Ankita Kapratwar et al. [5]: Static and Dynamic Analysis of Android Malware: They discussed a machine learning approach for the Android malware detection, performed system calls and permissions analysis, and faced challenges in dynamic features where they collected malware sample using MonkeyRunner; but this method may fail to execute the malicious parts of the code.
- Rosmalissa Jusoh et al. [6]: Malware Detection Using Static Analysis in Android: A Review of Features, Classification, and Obfuscation (FeCO): They compared all three types of analysis and used SAMAdroid for comparison of all analyses. Instead of that, we used a combination of DL with NCD and MDM.
- Belal Amro et al. [7]: Malware Detection Techniques for Mobile Device: They included three types of techniques: 1. stealth malware techniques, 2.

malware evasion techniques, and 3. malware detection techniques. In static analysis, they used permission-and signature-based analyses. However, in dynamic analysis, the permission-based approach is fast when we scan something but it may not work properly when we analyze something.

- Ruitao Feng, Sen Chen, Xiaofei Xie, Lei Ma, Guozhu Meng, Yang Liu, ShangWei Lin (2019) [8]: MobiDroid: A Performance-Sensitive Malware Detection System on Mobile Platform: MobiDroid is a nothing but a device that can tell us how much sensitive our malware behaves after the execution. It has a limited dataset for analysis, so we cannot rely on it.
- Matthew Leeds, Miclain Keeler, Travis Atkison (2017) [9]: A Comparison of Features for Android Malware Detection: In this analysis, they used a set of a Python scripts that can automatically analyze the data of the sample. The process control is based on reconnaissance. Here they limited fully utilizing resources; we could combine all datasets and train them together instead of training each one separately.
- Nirmala Yadav, Aditi Sharma, Amit Doegar (2016) [10]: A Survey on Android Malware Detection: In this research paper, they used executables to identify function calls in malware sample code. The disadvantage here is that they focused only on permission-based features.
- Hyunjae Kang, Jae-wook Jang, Aziz Mohaisen, and Huy Kang Kim (2015) [11]: Detecting and Classifying Android Malware using Static Analysis along with Creator Information: In this paper, they used an algorithm known as the detecting algorithm, which starts using the app's serial number mentioned in the certificate, and by this, we can by forget them. This method has a drawback in that we can only use this static analysis-based system.
- Vitor Monte Afonso · Matheus Favero de Amorim ·André Ricardo Abed Grégio Glauco Barroso Junquera ·Paulo Lício de Geus (2014) [12]: Identifying Android Malware Using Dynamically Obtained Features: In this research paper, they used API functions and system calls to classify whether the given sample or application is behaving maliciously. This method may fail during analysis when we do not have a simulation of the analysis environment.
- Yajin Zhou, Xuxian Jiang (2012) [13]: Dissecting Android Malware: Characterization and Evolution: They thoroughly analyzed the sample by identifying its characteristics, such as behavior breakdown, installation process, and activation of payloads. They also tried to develop next-generation anti-mobile malware solutions, which can help in the identification of IOCs.
- Ryo Sato1, Daiki Chiba2 and Shigeki Goto1 (2013) [14]: Detecting Android Malware by Analyzing Manifest: In this research paper, they research the static approach to Android malware analysis, which is based on the analysis of the Android Manifest file, which is in the APK source code. We cannot detect or analyze some samples with this method, which are packed by a packer.
- Niall McLaughlin, Jesus Martinez del Rincon, BooJoong Kang, Suleiman Yerima, Paul Miller, Sakir Sezer (2017) [15]: Deep Android Malware

Detection: In this paper, they discuss an Android malware detection system that uses a deep convolutional neural network (CNN) which can be used by malware detection systems. They don't use various types of datasets during transformation.

9.3 METHODOLOGY

This section discusses the setup of the malware testing environment, the proposed testing methodology, the different tools and applications used at each phase, and the different types of analysis performed. Throughout the process, time, ethics, and security were taken into account, considering the risks involved in malware analysis.

9.3.1 Testing Environment

An isolated virtual environment was configured on the system for the analysis. This setup was established to isolate the testing environment from the main system network. To set up the testing environment, we need one host system with the required specifications and two Android emulators for testing the malware samples. On the host system, we can install VirtualBox/VMWare, and in that, we need to install Santoku for analysis purposes. After that, we need to install Genymotion and add two Android emulator instances. Using ADB, we can connect those instances with Santoku and perform the analysis.

- Santoku Configuration: Santoku is installed in VirtualBox/VMware on the Windows Operating System, which has a network adapter configured as "NAT."
- Genymotion Configuration: Genymotion is installed on the Windows Operating System with two Android emulator in it for analysis purposes.
  Android Emulator -1: Android Emulator-1 installed in Genymotion with malicious APK installed in it for analysis purposes.
  Android Emulator -2: Android Emulator-2 installed in Genymotion with malicious APK installed in it for testing purposes.

9.3.2 Malware Analysis Methodology

As the use of Android applications increases, it attracts malicious threat actors. Hence, the security of Android devices has also become a significant issue. Analysts have come up with two main methods or techniques to analyze APK files to determine

TABLE 9.1
Tools and Technology Required

Item	Description	Unit
Work station (with Santoku installed)	16 GB RAM, i5 500 GB storage	1
Virtual device (Android emulator)	3 GB RAM, 20 GB storage	2
Virtual box installed with Ubuntu or Kali	4 GB RAM, 40 GB storage	1

whether they are malicious or not. The two main known techniques are static and dynamic methods.

- Static Analysis: Static analysis is a method of analysis that analyzes malware sample code and the full activities in an application within no range bound of time, without executing the application. The most important step in this analysis is the reverse engineering process, which can extract the code. We can also identify unknown malware using this analysis's machine learning approach.
- Dynamic Analysis: Dynamic analysis is a method that analyzes malware samples by executing them and running the APK to monitor and differentiate their behavior of the sample. For this, we need a good amount of resources and it can be costly. In dynamic analysis, some malware triggers at some point in time when the attacker wants to attack. For dynamic analysis, we need a separate environment for executing the malware sample, such as a sandbox; otherwise, it can affect other host machines. Therefore, we perform all these processes in an isolated environment.

We have used both static and dynamic analysis methods for testing Android malware. We use Apktool to analyze the APK and extract the Android Manifest file. Then, we analyze the permissions requested by the APK. In addition, we can focus on other parts of the APK, such as API calls and the components it uses. We can also focus on other parts of the APK for analysis like API calls and the components that the APK is using. First, we installed the APK in the virtual emulator, and then we started the analysis. After extracting the files from the APK using Apktool, we had the Android Manifest file with the Apktool.yml file, which contains the version and SDK version information of the particular APK.

9.4 MALWARE ANALYSIS USING SANTOKU: IMPLEMENTATION

In this section, we will work through the process of actually setting up the lab, configuring the various systems involved in the analysis process, and installing all the packages and tools needed for the purpose of malware analysis.

9.4.1 Santoku

Santoku is an open-source mobile forensics platform by which we can perform many tasks on mobile devices, such as analyzing them and performing vulnerability assessments. Using Santoku, we can connect mobile devices because it has an in-built SDK installed. We will be setting up Santoku in VirtualBox on the Windows Operating System. To configure Santoku, we first need to download its vmbox file for VirtualBox and then install it with the proper specifications.

Note: It is best practice to keep the system configuration in a "Host Only" mode. To achieve this, we install two network adapters on the Linux system that will host Santoku. One of the network adapters is configured with a static IP address that will be on the same network as the Santoku guest machine, and the other will be configured for normal activities such as updating Santoku and downloading and installing the necessary packages. We have many tools for analysis built into it. For example, for reverse engineering, we have Androguard, JD-GUI, Smali, etc.

9.4.2 Installing and Configuring Genymotion

Genymotion is a VirtualBox mobile emulator. We can install virtual mobile devices based on our requirements. Here we have different levels of APIs, based on which we can install the emulators. We are using Genymotion version 3.2.1 with two mobile devices installed in it. Genymotion allows us to create devices as per our own requirements, as well as provides already-built mobile emulators.

We have created two Android devices on which we performed the analysis. We installed the malicious app on these emulator devices and then performed the testing on them. We installed two devices:

1. Google Pixel with 9.0 API 28 version
2. Google Nexus 5 with 6.0API 23 version.

Now we are ready with the lab setup for testing. Now, we have to download the malware samples for analysis purposes.

Firstly, we analyzed malware sample using the static analysis method. From this, we obtained details about the permissions requested by the APK and the API calls it makes to the system. I extract the files from Apk using this Command

- apktool d (APK Name)
- cd APK Name
- cat AndroidManifest.xml

After that, we checked for other parameters to see if this APK is connecting with any external IP after executing it or if it is writing anything on external storage. We found some APKs that ask for external storage and Internet. Some of the APKs have Internet permissions, so we can say that after executing that application, it connects with some server and probably transfers our data to that server. On the other hand, some of the APKs have access to write on the external device, so if a user is using any peripheral device attached to the Android device, there is a chance that the attached device could also get corrupted or damaged.

We installed all 10 APKs into the Android emulator and then performed analysis on them. We used ADB commands to install the APKs through Santoku to the emulator. The Android emulator should be in host-only mode for a safe environment; otherwise, when we execute the APK, it could affect the host machine.

9.4.3 Static Analysis with VirusTotal

As we all know, in analysis, we cannot rely on only one tool or method. Therefore, for cross-verification and surety that the file is behaving maliciously, I used the VirusTotal platform. VirusTotal is an online service that analyzes malicious files and URLs to detect types of malware and malicious content using antivirus signatures and website scanners available in its database. We can upload our APK file and check whether it matches any existing malware signature. VirusTotal analyzes the APK file, matches it with every signature stored in its database, and gives us the best matching result.

9.4.4 Dynamic Analysis with Cuckoo Sandbox

As we know, for dynamic analysis, we need a separate environment so that the uploaded malware sample cannot affect the host system. For that, we configure Cuckoo Sandbox, which we can be used for the analysis of the APK sample.

Configuration & Installation of Cuckoo Sandbox:

- Cuckoo Installation: First of all, we installed Kali Linux in VirtualBox, and then in that distribution, we installed Cuckoo Sandbox as per the requirements.
- Then we installed the required libraries and frameworks using the Linux terminal, and after installing Cuckoo Sandbox, we needed to make changes to its config file.
- Therefore, for that, we need to use some config file commands to modify the config file based on our requirements and try to implement the filters accordingly.
- After making the config file changes, finally we can start Cuckoo Sandbox; so for that, first we have to start it from the terminal and then go to the browser and redirect to this URL: http://localhost:8000/
- Now start Cuckoo Sandbox by typing this command in the terminal: Cuckoo web
- Now redirect to the given URL in the browser: http://localhost:8000/
- After that, we can see the lot of processes going on in the terminal and framework starting up, and we can see the output to the given URL and observe the first look of the Cuckoo Sandbox environment.
- We can see the dashboard of Cuckoo Sandbox. Some other functionalities are also available, by which we can analyze our file in this isolated environment.

9.4.5 Analysis with VirusTotal

- VirusTotal provides facilities to analyze malicious APKs with existing signatures in their database, so we can match our APK with that and know the behavior of the APK we are using for analysis.
- Other than that, we can also track the activities of the malicious APK in VirusTotal through the analysis facilities provided by them. By examining the APK's connections with internal parameters, we have to know about its links with other targets.
- After that, we can know about the functions that will be used by the APK when we execute it and where it cause damage than we get to know about and we sometimes also get some network activities information that follow the APK after execution.
- Therefore, this way we can perform dynamic analysis with VirusTotal and extract as much information as possible from it related to the malicious APK file.

9.5 RESULTS

As per Table 9.2, we can say that the APK demands unnecessary permissions, which leads it to be considered malicious. We also checked for other versions of Android, but the results are the same for all versions from **version 6** to **version 9**.

As future work, the proposed approach may be enhanced by source code analysis to interpret the real intentions of API calls using sandboxing (dynamic analysis) and towards the real-time analysis of the Android device.

9.6 CONCLUSION

Static analysis techniques use static resources, which are available before the installation and execution of Android applications. As these resources contain all the resources the applications use to perform their actions, techniques based on static analysis are very effective in terms of malware detection. In static analysis, we only got the static details about the sample like the code and the resources it is using.

Dynamic analysis techniques use dynamic resources, which are available after the installation and execution of the Android APK sample, which we are using for analysis. By using this technique, we can extract each small piece of data from the APK and analyze everything we want from the APK like its behavior and characteristics. This analysis helps us to see the assembly-level function of the APK and gives us with details about it.

Therefore, by observing the results, we can conclude that dynamic analysis is somewhat more effective than static analysis because in static analysis, we are bounded with limitations, whereas in dynamic approach, all types of information about the uploaded sample APK are available. We got information about how it is going to behave after execution and how much it is going to create damage on the mobile device. From dynamic analysis, we can get real-time data on malicious APKs so that we can say that if this file or APK is behaving maliciously. In this way, we

TABLE 9.2
Result Table for Analysis of the Applications

Sample Name	Manifest.xml	Request for Internet	Download Extra Files	Request External IP Connection
com.bl.smsflash.apk	Yes	No	No	No
Dropdailer.apk	Yes	Yes	Yes	No
JiFake.A.apk	Yes	Yes	No	No
fakeAV_LableReader.apk	Yes	Yes	Yes	No
com.cattss.apk	Yes	Yes	Yes	No
com.chistescortos.apk	Yes	Yes	Yes	No
com.funnyys.apk	Yes	Yes	Yes	No
com.imagepets.apk	Yes	Yes	Yes	No
com.kitchenn.apk	Yes	Yes	Yes	No
com.laughtter.apk	Yes	Yes	Yes	No

can determine the IOCs from compromised Android devices using Android malware analysis.

Acknowledgment: This work was presented at the International Seminar on Emerging Technologies, Law, and Governance, sponsored by ICSSR.

REFERENCES

1. Belous, A., Saladukha, V., Belous, A., & Saladukha, V. (2020). Computer viruses, malicious logic, and spyware. *Viruses, Hardware and Software Trojans: Attacks and Countermeasures*, 101–207.
2. Jacob, G., Debar, H., & Filiol, E. (2008). Behavioral detection of malware: From a survey towards an established taxonomy. *Journal in computer Virology*, *4*, 251–266.
3. Sihwail, R., Omar, K., & Ariffin, K. Z. (2018). A survey on malware analysis techniques: Static, dynamic, hybrid and memory analysis. *Int. J. Adv. Sci. Eng. Inf. Technol*, *8*(4–2), 1662–1671.
4. Bakour, K., Ünver, H. M., & Ghanem, R. (2019). The Android malware detection systems between hope and reality. *SN Applied Sciences*, *1*, 1–42.
5. Kapratwar, A., Di Troia, F., & Stamp, M. (2017, February). Static and dynamic analysis of android malware. In *International Workshop on FORmal methods for Security Engineering* (Vol. 2, pp. 653–662). SciTePress.
6. Jusoh, R., Firdaus, A., Anwar, S., Osman, M. Z., Darmawan, M. F., & Ab Razak, M. F. (2021). Malware detection using static analysis in Android: A review of FeCO (features, classification, and obfuscation). *PeerJ Computer Science*, *7*, e522.
7. Moreb, M., Salah, S., & Amro, B. (2024). A novel framework for mobile forensics investigation process. *International Journal of Computing and Digital Systems*, *16*(1), 125–136.
8. Feng, R., Chen, S., Xie, X., Ma, L., Meng, G., Liu, Y., & Lin, S. W. (2019, November). Mobidroid: A performance-sensitive malware detection system on mobile platform. In *2019 24th International Conference on Engineering of Complex Computer Systems (ICECCS)* (pp. 61–70). IEEE.
9. Leeds, M., Keffeler, M., & Atkison, T. (2017, April). A comparison of features for android malware detection. In *Proceedings of the SouthEast Conference* (pp. 63–68). https://doi.org/10.1145/3077286.3077288
10. Yadav, N., Sharma, A., & Doegar, A. (2016). A survey on Android malware detection. *vol*, *7*, 47–53.
11. Kang, H., Jang, J. W., Mohaisen, A., & Kim, H. K. (2015). Detecting and classifying android malware using static analysis along with creator information. *International Journal of Distributed Sensor Networks*, *11*(6), 479174.
12. Afonso, V. M., de Amorim, M. F., Grégio, A. R. A., Junquera, G. B., & de Geus, P. L. (2015). Identifying Android malware using dynamically obtained features. *Journal of Computer Virology and Hacking Techniques*, *11*, 9–17. https://doi.org/10.1145/3077286.3077288
13. Zhou, Y., & Jiang, X. (2012, May). Dissecting android malware: Characterization and evolution. In *2012 IEEE Symposium on Security and Privacy* (pp. 95–109). IEEE.
14. Sato, R., Chiba, D., & Goto, S. (2013). Detecting android malware by analyzing manifest files. *Proceedings of the Asia-Pacific Advanced Network*, *36*(23–31), 17.
15. McLaughlin, N., Martinez del Rincon, J., Kang, B., Yerima, S., Miller, P., Sezer, S., . . . Joon Ahn, G. (2017, March). Deep android malware detection. In *Proceedings of the Seventh ACM on Conference on Data and Application Security and Privacy* (pp. 301–308). https://doi.org/10.1145/3029806.3029823

10 Technology Tools for Legal System

Understanding the Gaps and Enhancing the Capacity of the Stakeholders

Kapil Saini, Manmeet Dhillon, and Ravi Saini

"Embracing technology in the legal field is not just a choice; it's a necessity to stay competitive and provide effective legal services in the digital age."
– Joshua Lenon

10.1 OVERVIEW OF OUR LEGAL SYSTEM

The nation's legal system serves as the cornerstone for its society and government, facilitating conflict resolution and safeguarding individual rights while also promoting equality and impartiality for everyone. Moreover, a country's stability, social well-being, and prosperity are vital for a vigorous legal system. Technology is the great equalizer [1]. In this rapidly evolving world, it offers a lot such as case management, document analysis, and innovative solutions for legal research. Freeing up valuable time for legal professionals, solving complex legal problems, and automating repetitive tasks are enabled by technology. Moreover, large volumes of legal data also facilitate efficient management with the help of technology, adhering towards more informed decision-making and better results while empowering stakeholders so that they can work smarter [2]. This chapter titled "Technology Tools for Legal System: Understanding the Gaps and Enhancing the Capacity of Stakeholders" is concerned with understanding the power of technology in transforming the legal system, making it more accessible, affordable, and efficient for all. By analyzing the latest transformations and emerging trends in technology tools for the legal system, this chapter helps readers navigate the evolving landscape of legal technology effectively. Including the legal profession, technology has brought revolutionary changes across diverse sectors and completely transformed the way earlier legal professionals worked. When exploring the impact of technology, it is also crucial to study the functioning of the Indian legal system because the legal profession in India works within a complex framework where advocates, dressed in white and black, handle many files, represent parties with arguments, and judges who act as neutral arbiters

DOI: 10.1201/9781003512820-10

trying to balance between two parties without taking part in the forensic process. Furthermore, these files showcase the hard work done by lawyers, who invest much time and expertise in the critical analysis of evidence, extensive research, immense effort, and meticulous drafting skills by analyzing case precedents and assessing relevant facts and evidence for the preparation of files. Furthermore, court proceedings involve a series of steps, including filing petitions or complaints, issuing summons, pre-trial conferences, presenting evidence, examination of witnesses, oral arguments, and final judgment, which further depend on legal precedents. Every step involves procedural rules to ensure a fair and transparent legal process. Apart from lawyers and judges, courts rely on the collaboration of numerous other individuals to ensure smooth functioning. The following section introduces these stakeholders, emphasizing their vital roles and specific requirements.

10.2 UNDERSTANDING THE STAKEHOLDERS' NEEDS AND CHALLENGES

As mentioned above, various stakeholders play essential roles in upholding justice and maintaining the rule of law. Understanding the diverse needs of these legal stakeholders is crucial for developing and implementing technology-driven solutions that enhance efficiency, transparency, and access to justice within the legal field (Figure 10.1). Some of these stakeholders are listed below.

10.2.1 Judges

Judges hold the utmost responsibility as the principal decision-makers within the legal domain. Their pivotal role involves unbiased interpretation and application of the law to settle disputes and deliver judgments. They preside over court proceedings and ensure that due process is followed. They must possess a strong understanding of legal principles, fairness, and effective case management skills, for which they require access to comprehensive legal databases for research, efficient case management tools, and clear communication channels with court staff and legal professionals to fulfill their duties effectively. Coping with heavy caseloads, time constraints, diverse cases, and upholding impartiality presents formidable challenges for judges.

10.2.2 Lawyers

Lawyers are the reflection of their clients, representing their legal matters with their rights and interests. Efficient legal research tools, safe platforms for communication, and effective case management systems are used for solving complex legal procedures. Starting from conducting research to negotiating settlements, arguing, and presenting the case in front of the judge, lawyers are responsible for all. To manage their workload effectively, they require reliable time management solutions. However, lawyers face challenging caseloads with time constraints, so they must manage their workload effectively with reliable time management tools.

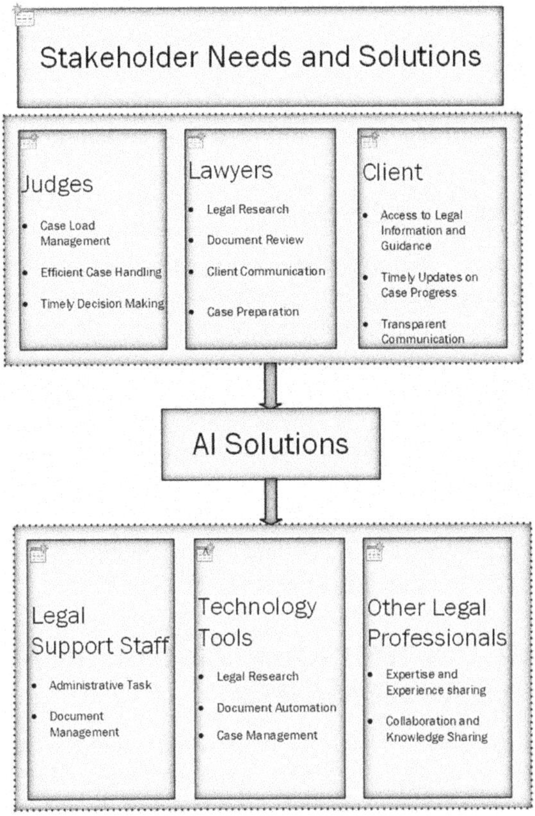

FIGURE 10.1 Stakeholder needs and solutions.

10.2.3 Litigants and Parties

Prompt, cost-effective, and transparent legal services are sought by individuals, businesses, or organizations because they want easy access, updates, and clear, open communication with their lawyers. They frequently need guidance through complex legal procedures and potential outcomes, and these needs vary widely, from dispute resolution to obtaining legal protection. Ethical and proficient legal representation is expected to meet their unique situations. For clients, the key points towards their cases are timely and cost-effective services with protected sensitive information, clear communication, and easy access to case updates. The lawyer's role becomes more crucial, especially when the client has less knowledge; when a lawyer sincerely handles the legal proceedings, then it can be overwhelming for clients.

10.2.4 Legal Support Staff

For the preparation of cases, communication with client, documentation and scheduling includes legal assistants, secretaries, and administrative personnel, and these play a vital role in aiding lawyers with all these tasks, because for the smooth functioning, all these things are important. Legal support staff benefit from effective tools for collaboration and safe data management systems to optimize their support services. However, managing a lot of paperwork, coordinating with lawyers, and multitasking can be more challenging for these dedicated professionals.

10.2.5 Law Enforcement Personnel

To perform their duties effectively, law enforcement personnel require real-time access to reproving information, such as custody records and warrants.

It is clear from the evidence that poor physical facilities, outdated technology, insufficient capacity of courtroom, and outdated infrastructure are reasons for stakeholders to feel burdened. Manual filing, record-keeping, and document management consume considerable time, are prone to misplacement, and make information retrieval cumbersome, leading to delays and inefficiencies. The challenges faced by them can be overcome with the help of improving technological advancements like digitized systems and electronic filing, which streamline legal proceedings. Additionally, less paperwork builds efficiency in managing crucial information, and building modern infrastructure also improves overall efficiency.

10.3 IMPERATIVE OF TECHNOLOGY INTEGRATION IN THE INDIAN LEGAL SYSTEM

This section fills the gaps between the legal system and the latest technological improvements like AI, machine learning, data analytics, and automation, providing solutions for improving the capacity of stakeholders. In addition to this, we will explore technologies that are promising for the future but not yet widely adopted. With legal technology, the legal sector showcases a revolutionary transformation through technology, representing unique tools and solutions. To empower legal professionals and enhance their efficiency and access to justice, legal tech incorporates a wide range of technology tools specifically designed for the legal field [3]. This section delves into the exploration of some of these technology tools, including AI-powered systems, data analytics platforms, document automation software, electronic discovery tools, and online legal research platforms, among others, examining their extent of usage and discussing the potential for further implementation.

10.3.1 AI-Powered Systems

In the Indian legal system, artificial intelligence (AI) demonstrates powerful potential. AI-powered contract analysis tools save lawyers valuable time by efficiently extracting and categorizing information from contracts, identifying crucial clauses, and highlighting potential risks. These software manage the creation, negotiation, and management of contracts, improve collaboration, ensure compliance, and automate contract workflows. Moreover, in the legal department and corporates, it is increasingly used and still its full potential is not yet realized; this can be done with further integration and customization opportunities. Historical data from similar cases can be used with the help of AI, which helps in predicting the outcomes and also this will help in finding out valuable insights into the strengths and weaknesses of their cases and in making more informed strategic decisions. Traditional methods of library-based research are being replaced by AI-powered legal research platforms such as ROSS Intelligence and LexisNexis's Lexis Answers, because they have extensive databases for cases. These platforms use natural language processing (NLP) to understand legal queries and provide relevant and accurate results, significantly speeding up the research process for lawyers [4]. Legal queries, answers, and advice can be given by AI-powered virtual legal assistants, and they can also interact with clients. To improve customer service and support, these chatbots use natural language, which allows them to be integrated into law firm websites. In this manner these make legal resources more accessible to the general public [3]. To ensure trust and confidence in the client-lawyer relationship, encrypted and confidential communication platforms can be used between lawyers and their clients. Additionally, AI tools can perform sentiment inspection on social media and online platforms to measure public thinking and analyze how legal matters are perceived by the public in shaping their strategies and communications. Furthermore, translators tools can be used with AI-driven machine, which assist in translating legal documents and information into different languages, improving international legal communications and cooperation. As a result, the scope of our legal stakeholders is significantly enhanced. This integration of AI-powered systems in legal matters can lead to enhanced efficiency, cost savings, and quicker decision-making capabilities for legal professionals.

10.3.2 Data Analytics Platforms

Data analytics platforms have also become an important tool in the legal field, helping legal professionals tackle the capacity of data and improve their overall practice. These plans of action use modern data analysis techniques, including machine learning, data mining, and visualization, to process and interpret large volumes of legal information skillfully. These platforms can filter through huge databases of legal documents and judgments to identify relevant case law and examples, providing insights to legal professionals to strengthen their arguments, assess potential risks, and craft effective strategies. They can forecast the consequences of cases, predict judge behavior, and possible bargaining options just by analyzing historical data, which permits lawyers to better expect case developments and advise their clients accordingly. Furthermore, document review, categorization, and prioritization can be done with e-discovery tools, allowing for a quick response to sensitive information, saving time and resources during legal proceedings. By using these platforms, legal professionals can process composite legal contracts, take out and organize key terms and sections using these platforms. Data-driven intuition can help improve case assignments, workload distribution, and resource utilization. Turnup legal trends and patterns can be easily identified and permit lawyers to stay ahead of developments in their practice areas and make data-backed decisions. Additionally, it can check client feedback, case outcomes, and service delivery to measure client happiness. In other words, by using data analytics platforms, legal professionals can alter how they proceed towards legal work, making their exercise more efficient, effective, and client-centric. These platforms enable lawyers to harness the power of data, leading to an agile legal practice.

10.3.3 Document Automation Software and Case Management Systems

With predefined templates and inputs from users, document automation and customized legal document tools can be generated; this will help in drafting the documentation, reducing errors, and saving time for lawyers. Meanwhile, case management systems make it possible to do effective document management, automation in workflow, and collaboration among legal teams. In tracking the deadlines and organizing client database, these systems are widely employed and proved to be helpful for stakeholders.

10.3.4 Virtual Courtroom Technology, E-Filing Systems, and Client Portals

Mutual benefits can be taken by judges, lawyers, and clients when virtual courtrooms and remote hearing solutions are used by the court because this reduces the need for physical appearances, reduces time, and increases access to justice. Moreover, the implementation of e-filing of documents can make it simple, reduce paperwork, and increase the efficiency of court operations. It also updates the real-time status of cases, which enhances transparency. Furthermore, for billing information and easy access to case updates, client portals can be beneficial.

10.3.5 Online Dispute Resolution (ODR) Systems

Online dispute resolution (ODR) systems have gained traction in resolving legal disputes outside traditional courtrooms. Platforms like Modria and FairClaims offer efficient and accessible alternatives for dispute resolution, especially in the context of e-commerce and consumer disputes. These systems employ advanced algorithms to facilitate communication, negotiation, and resolution between parties. ODR has proven effective in reducing case backlogs, promoting access to justice, and minimizing costs for both parties involved [5]. While the aforementioned technologies have made significant inroads, several promising tools are not yet widely used but have the potential to shape the future of the legal field. For instance, tools like Blockchain and Predictive Analytics. Blockchain is more advanced than contract management systems. It has the potential to create transparent, immutable, and self-executing smart contracts that could streamline transactional processes, enhance security, and reduce the need for intermediaries. Its broader adoption is currently limited by technical complexities and regulatory considerations. As far as predictive analytics is concerned, its utilization in the legal field is in its early stages, but it holds significant potential for future integration. Emerging technologies such as virtual reality (VR) for courtroom simulations hold promise for reshaping legal practice in the future by providing immersive courtroom simulations, offering an interactive environment for legal training, moot court exercises, and witness preparation. VR has the potential to enhance legal education and refine courtroom advocacy skills, but its practical implementation in the legal field is still in its nascent stages. Embracing the digital era,

India has taken significant strides towards modernizing its legal system, aiming to create enhanced platforms for legal stakeholders. This transformative endeavor seeks to bolster efficiency and facilitate improved access to justice for all. The next section talks about the modernization of the Indian legal system.

10.4 MODERNIZATION OF THE INDIAN LEGAL SYSTEM

The Indian government and the judiciary have undertaken various initiatives to modernize the legal system [6] and enhance its efficiency and accessibility (Figure 10.2). Some of the key initiatives are listed below.

10.4.1 eCourts Project

The eCourts Project was launched in India in 2005 as part of the National Policy and Action Plan for the Implementation of Information and Communication Technology (ICT) in the Indian Judiciary to digitize court processes, establish computerized case management systems, enable online filing, and provide electronic access to case records. The COVID-19 pandemic accelerated the adoption of virtual courtrooms and videoconferencing in India. These technologies enabled remote hearings, reducing the need for physical presence in courts and facilitating access to justice. They dropped down geographical barriers and improved the efficiency of court proceedings. The online filing system allowed litigants and lawyers to file cases, submit documents, pay fees, and track case status electronically, and it also provided e-services such as e-stamps, e-summons, and online payment of fines, making legal processes much more convenient and accessible.

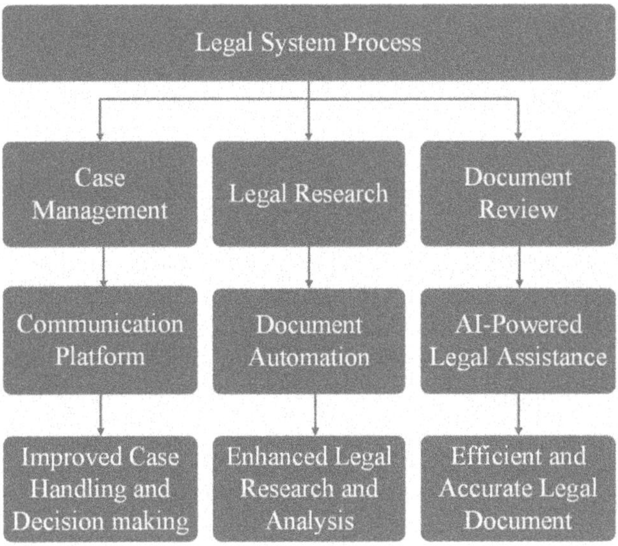

FIGURE 10.2 Legal system process.

10.4.2 Legal Information Systems

Efforts have been made to develop comprehensive legal information systems to provide access to legislation, case laws, legal precedents, and legal research resources online so as to help lawyers, judges, and researchers access legal knowledge quickly and facilitate informed decision-making. The National Judicial Data Grid (NJDG) is one such centralized online database. This promotes transparency and enables litigants, lawyers, and the public to access judicial information easily.

10.4.3 Case Management Software

Several legal institutions, including courts, have now incorporated case management software so as to automate case-related processes, track case progress, manage documents, and enable efficient communication among all stakeholders. These software solutions assist in streamlining court operations and reducing paperwork.

10.4.4 Virtual Legal Assistant Applications

As the name suggests, these applications provide assistance to legal professionals in their work. For instance, applications like Legal Bot Pro act as virtual legal assistants that are accessible round the clock, ready with all the legal information and advice, making it effortless for professionals and the general public. Other applications, such as Spot Draft and Kira Systems, help in analyzing, generating, and reviewing contracts, respectively. Applications like Case IQ and Vakalatnama assist lawyers and researchers in finding relevant case law and legal documents and with the filing legal documents and managing court appearances, respectively. The Legality application provides a platform that offers e-signature and document management with security and efficiency. Some other applications such as Lawrato and MyAdvo use technology to connect individuals and businesses with lawyers, providing access to legal services and resources, making it easier for users to find the right legal professional.

10.4.5 Legal Cloud Storage or Legal Document Management Systems

These platforms provide law firms, legal professionals, and organizations in the legal industry with secure and scalable solutions to store, manage, and access legal documents and data remotely through cloud technology. They offer features tailored to the unique needs and compliance requirements of the legal sector. The advancements in technology have indeed modernized certain aspects of the legal system, but as you know, legal technology is continuously evolving, and several challenges persist.

10.5 CHALLENGES IN ADOPTION AND IMPLEMENTATION OF TECHNOLOGY TOOLS

As mentioned above, the adoption and implementation of technology tools and systems have met with innumerable challenges. Some of these are listed below.

10.5.1 Digital Divide and Technological Illiteracy

Access to reliable Internet connectivity is still a challenge in many parts of India, leading to a digital divide in accessing legal services and limiting the potential benefits of technology in the legal system. Sometimes it is the connectivity; other times, it is the struggle to afford and integrate modern technology effectively [8]. Sometimes the pace at which technology advances also outpaces the ability of the legal system to effectively adapt and regulate, often creating a gap between technology and law. This widening digital divide and technological illiteracy is the biggest barrier to transforming the existing system into a more digitalized and modern legal system. Many times, individuals, particularly those from marginalized communities, may lack the necessary skills and resources to navigate digital platforms, file electronic documents, or understand complex legal technologies. Reducing this gap and building digital connectivity requires efforts to enhance digital literacy, provide training, and ensure equitable access to legal services.

10.5.2 Reliability and Accuracy

Undoubtedly, technology can streamline tasks, but at the same time, it can't be ignored that it can never match the creativity and judgment of the human mind. While technology can assist in legal research and analysis, ensuring the reliability and accuracy of the information provided by AI systems, it is not always devoid of errors or misinformation. This is also one of the reasons behind the reluctance to change established practices and potential disruptions.

10.5.3 Digital Security and Privacy Concerns

As we progress towards digitization, the protection of sensitive information and privacy becomes paramount [9]. Confidential information is always at risk of being compromised because of cybersecurity threats, data breaches, and hacking attempts. These issues erode public trust, hinder the legal system's integrity, and often lead to potential miscarriages of justice. Ensuring robust measures and staying ahead of emerging threats requires ongoing investments, technology upgrades, and the adoption of best practices, all of which can strain limited resources.

10.5.4 Big Data Management and E-Discovery

There are instances of delays, increased litigation costs, and potential omissions of crucial evidence. This may result in incomplete case presentations and erroneous judgments, affecting the credibility of the legal system because of the large volumes of unstructured data that need to be effectively managed so that it is easy to extract the relevant information as and when required without losing it.

10.5.5 Training and Capacity Building

Comprehensive training programs and capacity-building initiatives, including training on legal technology tools, data analytics, e-discovery, and other relevant areas, to fully leverage the benefits of technology in the legal domain, will enhance the digital skills of legal professionals and support staff.

10.5.6 Disinformation and Misinformation

The prevalence of misinformation and disinformation on digital platforms can adversely impact legal proceedings as these may taint jurors' perspectives, influence public opinion, and create an environment where the truth is harder to discern. Ultimately, this can lead to wrongful convictions, compromised investigations, and reduced trust in the legal system's ability to deliver justice.

10.5.7 Cybercrimes and Jurisdictional Challenges

Determining jurisdiction and apprehending perpetrators can be a complex issue when it comes to cybercrime [10]. Legal systems often struggle to address crimes that transcend international borders, leading to challenges in holding individuals or entities accountable for their actions, resulting in a perceived lack of accountability for cybercrimes and a sense of impunity among offenders.

10.5.8 Lack of Standardization

The lack of standardized formats or interoperability among different legal technology platforms makes it challenging to integrate and share information seamlessly. Thus, having applications to assist us is, in itself, a task, as even that requires synchronization and a level of standardization.

10.5.9 Artificial Intelligence and Automation

The implementation of artificial intelligence (AI) and automation technologies raises several concerns, such as biases in AI algorithms, translucent decision-making processes, and ethical considerations. Biases can emerge due to various factors, such as biased training data, algorithmic design choices, and the inherent biases of the developers, and these may lead to discriminatory outcomes in legal decisions. AI systems often operate as "black boxes," making it difficult to understand the decision-making process and the factors influencing outcomes. The lack of transparency raises ethical concerns, particularly in cases where AI is involved in critical legal decision-making [11]. To address these ethical concerns, stakeholders in the legal system must collaborate and establish guidelines and best practices. Ethical frameworks considering context-specific challenges should be developed to ensure transparency, accountability, and fairness in the design and deployment of AI technologies in the legal domain. These frameworks should consider striking the right balance between human judgment and technological assistance while addressing these concerns remains an ongoing issue for the legal system.

10.5.10 Regulatory Barriers

Last but not least, is the concern that relates to the legal implications that require appropriate regulatory frameworks [12, 13]. The development and use of technology tools in the legal system bring into the picture intellectual property issues that require our great attention, as it is important to protect the rights of all the parties involved. Before we talk about adherence to strict compliance with these rights, we need to have an effective legal framework to comply with in the first place. This is one of the major drawbacks. The technological advancements and a heavy reliance on them have raised questions of liability and accountability. In cases where technology is involved, it becomes difficult to determine whom to hold responsible, thus putting lawmakers in a challenging position. A clear, comprehensive, and robust legal framework is the need of the hour to ensure accountability for any adverse outcomes and also to deal with data protection and privacy concerns.

A solution to this is to make strategic investments in different spheres, such as our infrastructure, upgrading technology, initiating digital literacy programs, and enacting policy reforms so as to facilitate the seamless integration of technology into India's legal system.

10.6 SUCCESSFUL IMPLEMENTATION STRATEGIES

To ensure the effective implementation of technology tools into the legal system, it requires careful planning and adherence to the guidelines [14]. This subsection focuses on providing the guidelines for integrating technology tools into the legal system to maximize their benefits and mitigate potential challenges.

10.6.1 Developing Clear Objectives and Requirements

The first step towards successful implementation is assessing the specific needs of the legal system and identifying the areas where technology can enhance efficiency, accuracy, and accessibility. Stakeholders can align their efforts and select appropriate technology solutions by setting clear goals. Furthermore, to evaluate the suitability of different tools for the legal system, considering factors such as functionality, scalability, security, compatibility with existing infrastructure, and cost-effectiveness, a thorough technology assessment should be done.

10.6.2 Ensuring Data Privacy and Security

Data privacy and security should be prioritized for the integration of technology tools. To implement robust security measures, including encryption, access controls, and secure data storage, there should be collaboration between legal professionals and technology experts. Compliance with relevant privacy regulations, such as GDPR or CCPA, should be a fundamental consideration in the design and implementation of technology solutions [15].

10.6.3 Addressing Ethical and Legal Implications

The use of technology tools must follow ethical and legal rules. Stakeholders need to think about biases, fairness, and how it affects human rights. It is important to focus on transparency, accountability, and clarity to make sure technology is used fairly and ethically in the legal system. Policymakers should create clear rules to make sure technology tools meet ethical standards and safeguard personal privacy. Technology creators should focus on keeping data secure and being clear about their solutions.

10.6.4 Pilot Study and Continuous Evaluation

Developers must design algorithms well, think about various fairness measures, and conduct detailed audits and tests to reduce biases and support fairness. Bringing in people from diverse backgrounds can create more complete and fair tech tools. Technology tools need to be tested thoroughly and piloted before they are fully rolled out. Those involved should set up a way to gather user feedback and evaluate how the tools work in real situations. Ongoing evaluation makes room for updates and ensures that technology tools adapt to the changing needs of the legal system. Regular feedback between legal experts and tech professionals promotes updates in technology tools.

10.6.5 Training Programs for Legal Professionals

The effective use of technology in the legal system depends on the skills and knowledge of those involved. Therefore, it is very important to provide training and education to help stakeholders use technology well. These programs can include workshops, seminars, and online courses that cover topics such as legal technology, data analytics, and cybersecurity. Technology is rapidly evolving, and stakeholders must embrace a culture of continuous learning and adaptation. Legal professionals should actively seek opportunities to update their knowledge and skills through conferences, webinars, and professional development programs. By staying abreast of technological advancements, stakeholders can maximize the potential of technology tools in the legal system. Policymakers, legal professionals, and technology developers should prioritize technological literacy and training.

10.6.6 Collaborations and Other Initiatives by Institutions

To use technology well in the legal system, it is important that legal professionals work together with technology developers. Interdisciplinary collaboration, such as collaboration between legal institutions and educational institutions, can foster the development of specialized programs that address the intersection of law and technology and can lead to the development of more tailored and user-friendly technology tools [16]. By incorporating legal technology modules into law school curricula, future legal professionals can graduate with a solid foundation in utilizing technology tools in their practice. Regular communication and collaboration between legal professionals and technological capabilities can foster a shared understanding of goals, requirements, and constraints, enabling the development and implementation of technology tools that meet the specific needs of the legal system. This ensures that technology tools are designed with the end users in mind. Legal professionals can provide insights into the specific challenges they face, enabling technology experts to develop user-centric solutions that effectively address those challenges. Policymakers should facilitate partnerships between legal institutions and technology firms to co-create innovative solutions that address specific legal challenges.

10.7 SUMMARY

Technological disruptions are imminent to change the face of today's legal system by enhancing efficiency, accessibility, and accuracy. A legal practice overhaul is necessary, and these emerging recent trends, such as blockchain, natural language processing, and predictive analytics, offer promising opportunities for legal professionals to revamp their existing legal practices and improve the delivery of legal services. Policymakers, legal professionals, and technology developers must work together to navigate the challenges and make the best use of what technology has to offer. The ultimate takeaway is that a well-coordinated approach is required to transform into a modernized legal system that is agile, responsive, and accessible to all. By embracing the possibilities of technology while upholding core values, India's legal system can move forward into the digital age, fostering efficiency, fairness, and justice for everyone it serves.

REFERENCES

1. Leotti, S. M., & Slayter, E. (2022). Criminal legal systems and the disability community: An overview. *Social Sciences*, 11(6), 255. https://doi.org/10.3390/SOCSCI11060255.
2. Sil, R., Alpana, & Roy, A. (2021). A review on applications of artificial intelligence over Indian legal system. https://doi.org/10.1080/03772063.2021.1987343.
3. The Rise of Legal Technology. (2020, December 28). *Insight*. https://insight.mlaw.gov.sg/articles/future-of-law/2020-12-28-the-rise-of-legal-technology/.
4. The must-have legal technology tools for 2021. *Law Society Journal*. https://lsj.com.au/articles/the-must-have-legal-technology-tools-for-2021/.
5. CMS extends prior authorization for non-emergency ambulance trips. *JSTOR*. https://www.jstor.org/stable/27062786.
6. Dong, K., & Wei, N. (2023). Modernization of the basic public education service system. In *Transforming China's Public Services* (pp. 39–90). https://doi.org/10.1007/978-981-99-3943-5_3.
7. Urhibo, K., & Urhibo, K. (2023). The prospective nexus between constitutionalism, the rule of law, and democratic good governance: The Nigerian experience. *Beijing Law Review*, 14(2), 584–605. https://doi.org/10.4236/BLR.2023.142030.
8. Drossel, K., Eickelmann, B., & Vennemann, M. (2020). Schools overcoming the digital divide: In-depth analyses towards organizational resilience in the computer and information literacy domain. *Large Scale Assess Educ*, 8(1). https://doi.org/10.1186/s40536-020-00087-w.
9. Saura, J. R., Ribeiro-Soriano, D., & Palacios-Marqués, D. (2021). From user-generated data to data-driven innovation: A research agenda to understand user privacy in digital markets. *International Journal of Information Management*, 60, 102331. https://doi.org/10.1016/J.IJINFOMGT.2021.102331.
10. Dávid Tóth, Zsolt Gáspár., Jurisdictional Challenges of Cybercrime. (2020). *Journal of Eastern European Criminal Law*, 2, 101–118.
11. Grosz, B. J., et al. (2019). Embedded ethics. *Communications of the ACM*, 62(8), 54–61. https://doi.org/10.1145/3330794.

12. Whalen, R. (2022). Defining legal technology and its implications. *International Journal of Law and Information Technology*, 30(1), 47–67. https://doi.org/10.1093/IJLIT/EAAC005.
13. Regulation of legal technology in Singapore – issues, international scans, and ideas. *LawTech.Asia*. https://lawtech.asia/regulation-of-legal-technology-in-singapore-issues-international-scans-and-ideas/.
14. Machine bias: There's software used across the country to predict future criminals. And it's biased against blacks. *Benton Institute for Broadband & Society*. https://www.benton.org/headlines/machine-bias-theres-software-used-across-country-predict-future-criminals-and-its-biased.
15. Bias and ethical concerns in machine learning. *ISACA Journal*. https://www.isaca.org/resources/isaca-journal/issues/2022/volume-4/bias-and-ethical-concerns-in-machine-learning.
16. Doering, N., et al. (2022). Improving the relationships between indigenous rights holders and researchers in the arctic: An invitation for change in funding and collaboration. *Environmental Research Letters*, 17(6). https://doi.org/10.1088/1748-9326/ac72b5.

11 Future of Digital Forensics

Implications of Transparency and Accountability in Artificial Intelligence

Raunak Sood, Sarthak Sood, and Ranjit Karmahapatra

11.1 INTRODUCTION

Digital forensics (DF) is a crucial domain in the area of cybersecurity and law enforcement. A classical digital forensic investigation focuses on the identification, collection, preservation, analysis, and presentation of evidence in a court of law. The evolution of technology has led to a change in the dynamics of crimes. As crimes are taking place in the cyber world, there is a need for technologically advanced law enforcement agencies which can understand the dynamics of digital forensics. As the digital forensic field is dynamic due to regular updates, traditional forensic techniques are not effective. In this background, artificial intelligence (AI) has rapidly grown and offers solutions for conducting a revolution in the area of digital forensics. AI has become a transformative game changer that encloses a huge range of technology, including natural language processing (NLP), which can make forensic investigations more robust. The algorithms used in AI have a role to play in refining accuracy and effectiveness through the use of deep learning models (DLMs) to cut across complex datasets using artificial neural networks, hence increasing efficiency in tasks like image processing and speech recognition.

AI has the ability to process huge amounts of data, engage in pattern identification, and predict events, which may enhance the efficiency and accuracy of forensic investigations. The integration of AI into DF tools and workflows may help investigators automate routine tasks, increase the efficiency of evidence analysis, and uncover hidden evidence that may not be noticed by human experts during an investigation. AI can help in the increment of the precision of forensic investigations by minimizing the chances of human error, applying consistent analytical processes across the whole data structure, and reducing the risk of subject-based bias that might occur when data is examined by a human. In legal parlance, AI increases the accuracy of preserving the integrity of evidence, which can definitely change the course of

DOI: 10.1201/9781003512820-11

prosecution and antecedent investigation. Predictive analysis of historically collected data is incidental to the use of machine learning (ML) to see for any potential breaches or criminal activity before the occurrence of the actual incident. ML can help in studying trends found in the data, providing alerts about potential dangers, facilitating preventative actions, and improving overall security measures.

Integrating AI into DF raises a lot of concerns with respect to transparency, accountability, and the potential for bias as AI-based devices are becoming autonomous. This autonomy should not be misused by malicious actors; therefore, there is a demand for AI in its application should be fair, transparent, and unbiased. The future of digital forensics now rests on exploring the implications of transparency and accountability. An examination of the current state of AI integration in digital forensics connected benefits and challenges, must be addressed to harness the full potential of AI technology for the benefit of mankind. With automation being highly desirable nowadays, forensic investigations are facing a huge increase in data volumes, the variety of collected data, and extremely complex fields where only a few research fields offer a variety of solutions for overcoming the associated challenges. Knowledge representation of AI and semantic representation of forensic case data will help facilitate event correlation analysis, anomaly detection, timeline checking, and consistency checking.

The current limitations of AI in forensic processes that have been noticed are biases and lack of fairness, lack of transparency, and interpretability. Data privacy and security is a big limitation because of the mishandling of data and unauthorized access, which harms the integrity of data collected as evidence, thus harming the credibility of the whole forensic investigation. There are ethical and legal considerations in the connected usage of AI in digital forensics that are significant and lack ethical principles so that it does not infringe upon individual rights.

11.2 TRANSPARENCY IN ARTIFICIAL INTELLIGENCE-DRIVEN FORENSICS

The daunting task of defining transparency in AI can be done via a systematic set of mappings from the source of the object to mental operations. It means that metaphorically, when a human understands the structural aspects of knowledge, it follows a track of activities to give control over the activities of the algorithm and display it in the public domain to ensure the reliability of the content produced by AI [1]. Transparency is an indispensable condition for recognizing AI as trustworthy; the transparency and explainability of AI is a condition that has to be satisfied to protect personal data in situations of processing of data by AI [2]. When AI processes data, it takes and borrows data from open sources and then goes on to process data without even bothering about the copyright attached to the dataset being processed by it [3]. As open sources of data do not suffer from legal privacy law-induced storage limitations, AI goes on crunching data, which then leads to a compromising position regarding the liability and explainability of the content generated by the AI model [4]. According to Yuliya S. Kharitonova, there should be legislation that stipulates mandatory auditing of AI models, introduces transparency standards, and stipulates a compromise between technological abilities and advantages, accuracy, and explainability of results within civil law [5]. A legally mandated program should be introduced to certify the transparency of AI models; this certification process will solve the issue of liability of AI systems when used in miscreant-led activities [6].

Forensic processes often require an analytical approach to report categorical terms [7]. When AI is used in DF, one of the concerns regarding transparency in the decision-making process is maintaining trust and ensuring justice [8]. The major issue with categorical reporting is that there is an inherent lack of clarity; when the system simply categorizes evidence as "positive" or "negative" without attaching any kind of reasons, it increases the difficulty of understanding the decision-making process behind the workings of the AI-driven forensic algorithm [9]. Another notable problem with categorical reporting is that there is potential for bias, which may be either intentional or non-intentional to an unfair conclusion [10]. A need has grown for probabilistic reporting as it gives a more detailed view of the evidence by allotting different probabilities to outcomes [11]. This makes the decision-making process of AI-driven forensic tools more transparent and less biased [12]. Even though probabilistic reporting is complex for common people who belong to a non-technical background to develop a comprehensive understanding of the forensic report, hitherto probabilistic reporting gives a more accurate representation of evidence [13]. To bridge this gap, receiver operating characteristic (ROC) curves are being employed as they can give an accurate representation of the trade-off taking place between true positive results and false positive rates at multiple thresholds [14]. It is contended that by substantially using the ROC curves, it is observed that there is an increment in transparency in AI-driven forensic tools because it helps in the visualization of the decision-making process and shows different benchmarks showing different results. It helps in optimizing different benchmarks to balance sensitivity and specificity (accuracy) of the results, leading to pinpointed accurate conclusions. This may facilitate a better understanding of AI-driven forensic reports to be comprehensible by common people with non-technical knowledge. For effective implementation of ROC-facilitated reports to be treated as evidence in a court of law, it is necessary to train law enforcement officers in AI-driven forensic tools. The government can set up a nodal body to develop standards and guidelines for the usage of ROC curves. If possible, proficiency testing should be conducted to assess the analytical abilities of human controllers to check their ability to interpret and report evidence using ROC curves.

11.3 A GLOBAL PERSPECTIVE TO UNDERSTAND THE LEGAL CHALLENGES IN DIGITAL FORENSICS CONNECTED WITH ARTIFICIAL INTELLIGENCE

When AI is used in digital forensics, one of the key concerns is to ensure transparency and accountability within the decision-making process of the AI algorithm [15]. It is extremely important to maintain trust within the process and to make sure that the output is justice-oriented [16]. There are a lot of legal considerations arising with regard to transparency and accountability within the AI-driven forensic process [17]. The ROC curve analysis method offers a potential solution for enhancing clarity and making forensic evidence more reliable, especially keeping AI-driven decisions in a broader context [18]. There are some important issues that have been observed across various jurisdictions, and to get an overview of the developing global jurisprudential perspective, the steps taken by different jurisdictions are explained as follows:

1. United States: The admissibility of evidence is governed by the federal rules of evidence in federal courts [19]. However, the evidence that has been generated by AI has not been provided within the text of the federal rules of evidence; at the same time, the rules emphasize the reliability and relevance of the evidence being adduced in a court of law [20]. The 14th Amendment (Due Process Clause) requires that fair procedures be carried out when the accused is being addressed with criminal charges [21]. It is argued that any AI-driven decisions should be explainable and transparent to ensure compliance with the due process clause.
2. European Union (EU): The General Data Protection Regulation (GDPR) places greater emphasis on explaining AI-automated decision-making. The Artificial Intelligence Act sets out specific compliance requirements for high-risk AI systems, including transparency, accountability, and human oversight [22]. It is argued that the EU model also promotes transparency in AI and seeks to promote transparency, thus orienting policy towards making AI transparent by its designed model.
3. China: China's cybersecurity law requires organizations to ensure the security and reliability of AI systems [23]. The law protecting personal information addresses privacy and security, including the use of AI in data processing [24]. It is contended that China is trying to orient its policy where AI-related transparency is being introduced in the data processing stage and not the manufacturing stage.
4. India: India has the Digital Personal Data Protection Act, 2023 (DPDP Act), which does not directly address AI, but it emphasizes data privacy and security, which are important for AI-driven systems and forensic abilities [25]. The Information Technology Act, 2000 (IT Act) governs cybersecurity and digital provenance [26]. It is relevant to use AI in digital forensics, especially in terms of data security and privacy [27]. It is contended that even though Indian law does not explicitly include AI transparency within the ambit of its law, there is an indirect consensus building within its legislation that fair use must be made of AI technology; thus, any "fair use" of AI should include "transparency" and "accountability" within the AI algorithm.
5. Australia: Australia has its own privacy act, which governs the collection, use, and disclosure of personal information of data subjects [28]. This law is relevant to the use of AI in digital forensics, especially highlighting the privacy and security issues involved during the collection of data [29]. Competition and consumer acts are relevant for AI in digital forensics, especially in terms of the accuracy and reliability of evidence that has been analyzed or generated by AI if it affects consumers or is used in any anti-competitive manner.
6. Canada: The Personal Information Protection and Electronic Documents Act (PIPEDA) is used in the governance of the collection, usage, and disclosure of personal information, which is relevant to AI in digital forensics, with greater respect to data privacy and security, thus highlighting the need for the use of transparent AI in digital forensics [30].

After conducting a thorough analysis of the laws of the countries mentioned above, the common legal challenges and the role of ROC curve analysis can be seen to help in the admissibility of AI-generated evidence. It comes with a caveat that courts must act as gatekeepers to assess the authenticity of AI-analyzed evidence, which includes the algorithm of an AI-driven forensic tool and the data prompt input into the tool by the forensic investigator. The transparency and explainability of AI systems must be present to ensure fairness and accountability, which the ROC

TABLE 11.1
Global Perspective of Key Issues and Potential Solutions Offered by ROC Curves

Sr. No	Country	Applicable Law	Key Issue	Solution Offered by ROC Analysis Curve
1.	United States	Federal Rules of Evidence, 14th Amendment (Due Process Clause)	Reliability of evidence, checking the relevance connected with the legal framework. Fair procedures connected with the right to confront accusers.	It shows reliability and accuracy connecting sufficiently with AI-driven decision-making inside the forensic tool.
2.	European Union	GDPR, AI Act	Right to seek an explanation for automated decision-making, introducing transparency, and keeping essential human oversight over the algorithmic process. Making persons more accountable.	ROC curves give a visual graphical representation of the whole decision-making process that has been taken by the AI-driven forensic tool.
3.	China	Cybersecurity law, Personal Information Protection Law	Seeks the security and reliability of AI systems. Data privacy and security are concerns connected to the legal framework.	Helps in the identification and mitigation of bias within the AI-driven decision-making forensic tool.
4.	India	DPDP Act, IT Act	Data privacy, cybersecurity, accuracy, and reliability of AI-generated evidence connected with the Bhartiya Sakshaya Adhiniyam, 2023	ROC curves allow for human oversight and assessment of the performance of AI.
5.	Australia	Privacy Act, Competition and Consumer Act	Data privacy, cybersecurity, accuracy, and reliability of AI-generated evidence	Aids in the presentation of admissibility in court of law with regard to AI-generated evidence.
6.	Canada	PIPEDA	Privacy and cybersecurity	ROC curves greatly contribute to transparency and allow easy presentation to judges, juries, and other justice system stakeholders.

curve analysis will be able to successfully supplement to the desired need of the digital forensic expert. ROC curve analysis will give a visual representation of the AI decision-making process, which will help make it easier for the judge, jury, and other stakeholders within the criminal justice system. The bias and discrimination that can be outright seen in AI systems might percolate into the output results. ROC curve analysis will help in the identification of these biases in AI-based forensic decision-making. Human oversight is essential, as seen in the EU AI Act and GDPR; ROC curve analysis will help give a framework for human oversight and allow digital forensic experts to assess the performance of AI systems, thus helping in making informed decisions. By resorting to a comprehensive address of these legal considerations and using ROC curve analysis, AI-powered digital forensics can be used in an ethical, responsible, and justice-oriented manner, thereby acting as a subset of the principle of fairness embedded within the algorithmic watermark of the ROC-based AI analysis curve.

11.4 ANALYZING THE IMPACT OF LACK OF TRANSPARENCY IN AI-DRIVEN FORENSICS: A CASE STUDY OF IMAGE ANALYZER

A case study of image analyzers shows the potential risks attached to the lack of transparency in AI-driven forensic tools [31]. Though such tools may be extremely result-oriented and powerful for acting as an aid in the investigation of complex cases, this might lead to significant issues [32]. One of the issues can be a lack of understanding of the decision-making process because of the usage of opaque algorithms; the algorithms used by Image Analyzer to categorize different images are not fully displayed in the forensic report. This lack of transparency might cause, much hindrances in determining how the AI-driven forensic tool came of this conclusion [33].

TABLE 11.2
Impact of Transparency in AI-Driven Forensics

Sr. No	Issue	Impact	Proposed Solution
1.	There is a lack of understanding of the decision-making process.	This leads to increased difficulty in determining accuracy and reliability, hence increasing the potential for biased outcomes.	Model interpretability techniques like feature importance analysis and audits and evaluations are some of the potential answers.
2.	Difficulty in identification and mitigation of errors	Limitations of error insights and many vulnerabilities are being shown towards adversarial DF.	Improvements can be made in error reporting and analysis. Officials can be trained, and model design can be made more robust.
3.	Impact on forensic investigation	Misleading evidence, justice being delayed, and losing public trust.	Human oversight and validation, along with monitoring and improvements, are among the potential solutions.

Another set of issues that might crop up during the latter stages is potential bias and questionable accuracy because the AI system might be unintentionally biased, leading to potentially harmful conclusions [34]. Furthermore, another observed problem is the difficulty faced in identifying errors and mitigating any disruption that might be caused by potential errors. The error codes given by the system give limited information about the reason behind the error; hence, it is extremely difficult to engage in identifying the error, and it is possible that adversaries may exploit this lack of transparency in Image Analyzer to manipulate system output, thus leading to false positives or negatives [35]. The impact on forensic investigations is that there is a lot of misleading evidence that is inaccurate or biased in its classifications, which can compromise the investigation as a whole. The major consequence is that there might be significant delays in the investigation, leading to a significant erosion of public trust [36]. The potential solutions include, but are not limited to, model interpretability, which in itself is furthering the development of techniques that make the decision-making process more transparent. Using features like importance analysis or visualizing the decision boundaries can help in making the algorithm lean towards fairness. There is a need to conduct regular audits to assess the accuracy and fairness of the system output [37]. The model needs to be trained on adversarial examples to improve robustness against attacks [38]. The AI system needs to be continuously monitored for signs of attacks, and if attack signs are visible, appropriate countermeasures need to be taken. Human oversight is the best mechanism alongside feedback loops [39].

11.5 ACCOUNTABILITY IN ARTIFICIAL INTELLIGENCE-DRIVEN DIGITAL FORENSICS

When AI is integrated into forensic tools, any decision made on behalf of the AI-enabled forensic tool is bound to face challenges in relation to its accountability, thereby complementing the demand for careful examination and regulation [40]. The issues seen in AI-driven forensics are manifold and extremely critical to address so that a fair investigation can occur. The major challenge is that sometimes AI algorithms function as opaque black boxes, with decision-making not being accessible to human beings [41]. This amount of opacity then causes hindrance in the ability of the AI tool to further analyze and understand specific conclusions like how the risk scores are being reached [42]. There are cases like COMPAS and UKVI where the absence of transparency may lead to complications for public oversight and cause issues for AI-driven forensics to be accountable [43].

Furthermore, there are similar challenges like hidden biases within the algorithm, which are not discoverable even during ordinary due diligence carried out by forensic experts. These biases within the algorithm may affect the decision-making powers of individuals based on factors like nationality, ethnicity, age, sex, or past records, thus compromising the fairness and objectivity expected in forensic reporting and decision-making [44].

For determining liability in AI-assisted forensics, it is proposed that there should be shared responsibility between the AI developers and users because developers may create algorithms that influence AI-generated outcomes. Therefore, the complex interaction between human decision-making users and the AI algorithm will blur the line of accountability [45]. End users like forensic experts and government officials, will still need to make the teleological final decisions generated by the AI-driven forensic model [46]. Thus, regulation and policy should be oriented towards the end goal of determining the legal standards of responsibility sharing between both the AI developers and users of AI [47]. Liability should be developed keeping in mind the standard of a complex interplay of human and AI decisions; regulations should be targeted to ensure that human insight prevails over the operational part of the AI algorithm [48]. Human decisions should be able to override any possible algorithmic biases and inaccurate input.

Another issue that has cropped up is the need to conduct an impact assessment of the impact of data decontextualization and functional simplification. Wherever possible, data fragmentation and recontextualization should be done, as pointed out by leading scholars that AI algorithms have been observed to give different meanings to data input, thus stripping the whole meaning out of the context of the original input data [49]. Then, out-of-context stripped data is assembled by the AI algorithm according to its requirements and needs [50]. This process risks decontextualization and simplification, dehumanizing the human input by focusing on particular data points like age, nationality, and ethnicity without taking into consideration other aspects of human life [51]. Functional simplification allows algorithms to operate in an efficient manner but distorts original data connected to complexities. An example of the COMPAS recidivism algorithm can be cited, where the risk score calculation, which relies on different factors, can be addressed, thus not accurately representing the individual's true risk potential.

Some of the legal and ethical points for accountability that are emerging include the right to transparency and accountability in AI algorithms connected to the maintenance of the delicate balance between intellectual property and public transparency. In cases where decisions are taken by public sector authorities, the ethical and legal implications state that citizens should be allowed to understand the working mechanism behind the AI-taken decision [52]. For example, if the public is unable to access or engage in the scrutiny of the AI process that deals with visa approvals or sentencing in court, it will raise concerns about the lack of accountability. Even though proprietary rights are protecting some parts or specific mechanisms within the AI system, there is a need for balancing when the system is used in public-related rights, liabilities, and duties connected to decision-making [53]. Public institutions using AI in their decision-making processes should take more accountability measures to mitigate the "black-box" risk-based algorithms affecting individual lives without proper explanation or remedy [54].

Some policy recommendations to alleviate the concerns highlighted above may possibly include introducing more explainability in AI design so that AI decisions are transparent to end users and the public. This could, towards the end, require AI models to give reasons for their outputs, or regulatory mandates could be placed so that models can be used in public services and thus be auditable [55]. For AI-driven forensics, the only suggestion would be that the government comes up with guidelines and due diligence on the part of various intermediaries to secure the AI environment and make its usage more trustworthy [56]. Any algorithmic decisions in forensics should incorporate the "human-in-the-loop" principle, which states that human insights and supervision should be present at two stages: firstly, during the input of commands into AI, and lastly, during the output stage where the AI is deciding on the rights, liability, or duty of a human being [57].

TABLE 11.3
Accountability and Policy in AI-Driven Forensics

Sr. No.	Key Issue	Description of Key Issue	Proposed Policy Recommendation
1.	Transparency and AI algorithm	It has been observed that AI algorithms sometimes lack transparency and function similarly to opaque black boxes, making their comprehension difficult when connected with the decision-making process, which leads to various hidden biases.	Implementation of measures where the AI decision-making explained whenever used in public or forensic applications and interfaces.
2.	Assigning responsibility	Responsibility assignment to a particular individual or developer is very challenging; hence, sharing between developers and end users relying on AI-generated outputs is something that needs to be worked on in depth.	One of the proposed solutions can be to define the scope and limits of accountability, emphasizing the developer and end user roles and responsibilities within the process connected with AI-assisted cases for forensic-related applications.
3.	Data decontextualization	Another area of concern is the recontextualization of data by AI because it engages in the simplification of complex information by creating different fragments of data points, which may not accurately represent the profile of an individual.	Preservation of data context whenever feasible, which will prioritize a holistic view of data and then help make the review of decision-making easier wherever possible.
4.	Oversimplification	Functional simplification of data can lead to distortions in real data-based complexity, which will reduce the people's data to an equivalent quantity of a "risk profile." The risk profile will then be based on factors with different weights attached to them.	It is proposed that regulations should mandate AI systems to allow for context-based and context-sensitive inputs, which will further help mitigate oversimplification in high-profile forensic cases.
5.	Balancing intellectual property and public access to AI	Proprietary algorithms present in decision-making raise concerns about transparency and public rights because the public will lack valuable insights into the workings of the algorithm.	It is better to tailor regulations in a way to ensure that standard operating procedures are made, which further the "human-in-the-loop" protocol, thus ensuring human review and ethical oversight.

The need for accountability in AI-assisted forensic decision-making has multiple facets that require both technical and policy-based transparency for the protection of the rights of various individuals. It is important to ensure that the AI system is held accountable, which will not only improve the model as a whole but also enlarge the scope of interpretability, thus shaping the regulatory standards for addressing the ethical implications of algorithmic decisions in the public sphere.

Try to add a case study on image categorization. Image categorization depends on the type of data fed into the AI system, which might be useful in the case of POCSO (Protection of Children from Sexual Offences) cases. India must create a database like Project VIC International (Microsoft's PhotoDNA tool); these kinds of tools can help law enforcement officers reduce the need to review near-duplicate images in a case.

11.6 INTERSECTION OF ETHICS AND LAW IN AI-ENABLED DIGITAL FORENSICS

The integration of AI into digital forensics leads to a whole new domain of ethical challenges across many stages of the investigative forensic process [58]. Some of these challenges have a major influence on the collection of evidence, analysis, and presentation of forensic reports in ways where the legal standards are unable to keep up with the connected ethical issues [59]. One of the foremost challenges is "inconclusive evidence" [60]. The key ethical challenge associated with inconclusive evidence is that when AI is applied to digital forensics, it leads to uncertain or unreliable evidence because AI algorithms often rely on statistical models for generating insights on data. Thus, the conclusive derived when AI is connected with digital forensics is that the evidence-based result is more probabilistic in nature rather than definitive [61]. Upon closer analysis of the forensic stage, the analysis and evaluation phase of the forensic process involves the generation of inconclusive evidence, which misguides forensic conclusions in determining the likelihood of involvement or intent [62]. The legal issue faced here is regarding the probabilistic evidence in court, which can undermine the reliability of forensic findings and may violate standards requiring clear and convincing evidence [63]. Another challenge of inscrutable evidence arises when the black-box nature of AI limits transparency and limits the scope of investigators, defense counsel, and judges from understanding how evidence was derived [64]. The forensic aspect involved here is that in the presentation phase, the inscrutable evidence will make it extremely difficult to explain in court the relevance described by AI in the forensic report because stakeholders might not be able to grasp the whole algorithmic process used in the forensic stage [65]. The legal implication is that the lack of transparency challenges the basic human right to a fair trial, and for defense counsel, it becomes difficult if the workings of the AI are not accessible. This further violates the principles of natural justice, as clarity and verifiability of evidence have been challenged at the very basic level.

Furthermore, another aspect of the evidence is misguided evidence (evidence without any direction when input into an AI system). Misguided evidence faces an ethical challenge where AI-based forensic conclusions are drawn by the algorithm attributable to the AI-powered forensic tool [66]. The result produced by the AI-powered forensic tool is reliable only to the extent as per the data input in the initial stages, leading to flawed conclusions by the AI system [67]. On a forensic level, this issue may arise during the collection of evidence and processing in a forensic laboratory, where faulty data may lead to inaccurate reconstructions of events or identities [68]. The legal complication is that poor quality of input commands can lead to a grave miscarriage of justice. If the conclusion drawn by the AI tool is rife with false positives, it will ultimately harm the evidentiary weight of the evidence sought to be analyzed in a court of law [69]. AI can inadvertently reinforce biases within the forensic decision-making process, leading to various discriminatory results based on the demographics of evidence adduced within the algorithm, which might unconsciously lead to the profiling of various individuals [70]. On a forensic level, disturbances may be caused at the profiling and risk assessment phases, where the AI may be able to flag individuals based on the data within its training datasets rather than actual behaviors [71]. The legal conclusion that may be drawn is that bias in forensic assessment may conflict with principles of fairness and equality before the law, risking violations of human rights as decisions are based on data that can wrongfully impact subclassified groups of people [72].

AI is having transformative effects on evidence which is another ethical challenge as AI can change the perception and framework of the evidence structure when integrated with a forensic tool [73]. AI-powered forensic tools can influence forensic evidence by embedding unintended language into the analysis report of the evidence. At the forensic stage, the use of AI can change investigative practices, but caution should be taken to not prioritize AI-generated insights over traditional forensic evidence. The legal implication is that there could be a shift in the introduction of new biases into the legal system regarding the creation of dependencies on AI at the expense of conventional methods [74]. Courts may need to consider adapting evidentiary standards to account for these evolving forensic tools and techniques. On an ethical level, AI decisions involve multiple stakeholders like developers, users, and organizations, which makes it difficult to assign responsibility if an error occurs [75]. The interconnectedness of these parties creates a complicated question of defining the limits of accountability. At the forensic stage, the handling and verification of evidence, if handled by AI, can cause traceability issues, thus making it hard to identify which agent of the AI may have introduced the fault, impacting the integrity of the chain of custody [76].

AI in digital forensics may rely more on correlations without any part of causation being affected, thus leading to actions or conclusions that aren't justified in totality [77]. For example, profiling based on the correlations between data may flag individuals without any form of concrete evidence. Forensically, the investigation may be harmed because AI may prompt unjustified searches based on correlational data; thus, a legal consideration of lack of causative backing may breach the requirement of "reasonable suspicion" under criminal law [78]. Courts may question the validity of AI-decided actions as they might lack justifiable grounds. AI algorithms may yield discriminatory outcomes by acting in a disproportionate manner targeting a set of demographics; hence, bias in forensic tools may lead to profiling or exclusion during the identification stage of the investigation [79]. Forensically, this might prove fatal to the project because the algorithm may target individuals from specific backgrounds based on biased criteria [80]. This discriminatory practice followed by

AI-based forensic tools and techniques could violate the equality clause and potentially increase the timeline of litigation, leading to various appeals or reversals if AI-based biases are identified in the forensic reports presented before the court.

Informational privacy is another dimension of ethical concern that needs to be looked at. AI-based profiling infringes on the privacy of individuals, often relying on sensitive data for the creation of profiles and drawing conclusions [81]. On a forensic level, privacy is relevant in the data collection phase, where the AI system might aggregate the personal data of individuals without procuring express consent from that particular individual [82]. The legal complication that might occur due to the unauthorized use of data may lead to violations of privacy law, where legal standards for the usage of data in forensic analysis need to be rigorously enforced to protect individual rights. Determination of responsibility in AI-taken actions is very difficult and complicated, especially when AI systems are operating independently [83]. On forensic grounds, challenges arise throughout the forensic process, especially when AI engages in significant contributions to decision-making. The law should be tailored in a way that there is a clear establishment of who bears the liability for errors, which might require the redefining of accountability in forensic and legal standards [84].

There are significant challenges of AI-driven forensic tools regarding safety and resilience to avoid malfunctioning because even the most minute error may undermine the investigation. On a forensic level, this applies across all stages, especially during evidence collection and analysis, where AI may be frequently used [85]. In such stages of forensic analysis, AI accuracy is extremely important. The legal complication is that there might be malfunctions in AI tools, thus leading to wrong analyses of evidence; hence, a legal mandate is necessary for AI reliability in forensic applications [86]. One of the proposed solutions to all ethical concerns can be carrying out ethical auditing within all prevailing AI-driven forensic tools. Systematic ethical auditing is useful in identifying and correcting ethical violations. Across all phases of the forensic investigation process, auditing can take place to make sure that AI functions adhere to ethical standards. On the other hand, law might be drafted in a way that mandates ethical auditing, thus becoming a regulatory requirement for maintaining the integrity of evidence and ensuring fairness within AI-driven forensics [87].

As for the ethical challenges associated with regulatory concerns, there is a need to connect ethics to legal and regulatory frameworks to ensure that AI systems are operating in a fair, transparent, and responsible manner across the whole forensic process. In addressing these concerns, help can be ensured that AI serves justice ethically and effectively.

11.7 POLICY RECOMMENDATIONS AND WAY FORWARD

The legal frameworks of countries across the globe should be tailored in a manner to enforce standards that mandatorily require AI-based evidence to include high levels of certainty. There should be a mandate for AI-based transparency within all regulations across AI algorithms being used for the generation of text. If possible, AI-by-design should be enabled to make its output more transparent and self-regulate the labeling of the content generated by it. Enforcement of strict data quality standards should be mandated along with the requirement of verification and validation of input sources of data across AI-driven forensic models. AI-driven forensic models should be ingrained with the principles of fairness; hence, to promote fairness, ethical audits must take place in AI profiling tools for the enforcement of non-discrimination policies. AI should be regularly evaluated to adjust for societal impact and legal standards as per advancements in it. There is a need to develop accountability frameworks that specify the responsibility of users and developers based on AI outcomes. There is a need for a cause-based approach to AI-based forensics so that the "reasonable suspicion" standard under the criminal justice system can be maintained. The law should mandate privacy compliance audits to enforce data usage policies for safeguarding personal information if legal frameworks could define the responsibility and liability in AI-assisted processes, including shared responsibility frameworks.

11.8 SCOPE FOR FUTURE WORK: AI-BASED DATA TRANSLATION

As far as the concept of AI-based data translation is concerned, there is a data transformation when being presented by an AI-powered forensic tool if asked to switch the language. It has been observed by DF experts that the evidence is present in the form of the French language. Now, when AI-powered forensic tools start reporting the same evidence in English, the words used by the AI lose the meaning of the words being conveyed in French. This is a serious problem faced by multiple forensic experts. This is a significant drawback while presenting evidence in a court of law.

11.9 CONCLUSION

AI-powered forensic tools hold a huge amount of potential for enhancing investigative accuracy and efficiency, but they come with their own set of risks. AI is having an increasing role in the analysis of evidence processing and analysis; thus, AI algorithms behind the making of forensic tools should be more transparent and legally compliant. International frameworks like GDPR and the EU's AI Act highlight the mandate for transparency and ethical use of AI. Forensic investigation has various stages; AI-powered forensic tools might unintentionally work towards the profiling and risk assessment of criminals. AI is shaping forensic practices and introducing unintended reforms and biases; thus, the legal and regulatory frameworks need to update their policies and establish a shared accountability model for assigning responsibility to developers and end users. Adoption of ROC curves for visualization of data within forensic reports will enhance data quality checks, along with support from judicial oversight, which allows courts to evaluate AI-derived evidence.

REFERENCES

1. Larsson, S., Heintz, F., Department of Technology and Society, Lund University, Sweden, & Department of Computer Science (IDA), Linköping University, Sweden. (2020). Transparency in artificial intelligence. In *Internet Policy Review* (Vol. 9, Issue 2). https://doi.org/10.14763/2020.2.1469
2. Kharitonova, Y. S. (2023). Legal means of providing the principle of transparency of the artificial intelligence. *Journal of Digital Technologies and Law*. https://cyberleninka.ru/article/n/legal-means-of-providing-the-principle-of-transparency-of-the-artificial-intelligence

3. Felzmann, H., Villaronga, E. F., Lutz, C., & Tamò-Larrieux, A. (2019). Transparency you can trust: Transparency requirements for artificial intelligence between legal norms and contextual concerns. *Big Data & Society*, *6*(1). https://doi.org/10.1177/2053951719860542
4. Digital Evidence Management – Transparency and Accountability. (2023, November 8). *CentralSquare*. https://www.centralsquare.com/resources/articles/digital-evidence-management-transparency-and-accountability/
5. Expert Code Lab Pvt. Ltd. (2023). Interactive forensic science hands-on events and workshops role of forensics in human rights investigations. *Learn Forensic*. https://www.sifs.in/events/blog-details/forensic-in-human-right-investigation-aloi-renz-santos
6. Team, E. (2024, September 16). *Ethics in Forensics*. https://edinbox.com/council/forensic-sciences-gfsec/2494-ethics-in-forensics
7. Catoggio, D., Bunford, J., Taylor, D., Wevers, G., Ballantyne, K., & Morgan, R. (2019). An introductory guide to evaluative reporting in forensic science. *Australian Journal of Forensic Sciences*, *51*(sup1), S247–S251. https://doi.org/10.1080/00450618.2019.1568560
8. Champod, C., Biedermann, A., Vuille, J., Forensic Science Ireland, Willis, S., & De Kinder, J. (2016). ENFSI guideline for evaluative reporting in forensic science, a primer for legal practitioners. *Criminal Law & Justice Weekly*, *180*, 189–193. https://serval.unil.ch/resource/serval:BIB_842A724DBA9E.P001/REF.pdf
9. Biedermann, A., Champod, C., & Willis, S. (2016). Development of European standards for evaluative reporting in forensic science. *The International Journal of Evidence & Proof*, *21*(1–2), 14–29. https://doi.org/10.1177/1365712716674796
10. Bonsu, D. O., Afoakwah, C. B., Abedi, M., Higgins, D., & Austin, J. J. (2022). Ethics reporting in forensic science research publications – a review. *Forensic Science International*, *335*, 111290. https://doi.org/10.1016/j.forsciint.2022.111290
11. Wu, Y., Ou, C., Chien, P., Chang, H., Ko, J., & Wang, B. (2018). Chidamide-induced ROS accumulation and miR-129-3p-dependent cell cycle arrest in non-small lung cancer cells. *Phytomedicine*, *56*, 94–102. https://doi.org/10.1016/j.phymed.2018.09.218
12. De Keijser, J., & Elffers, H. (2010). Understanding of forensic expert reports by judges, defense lawyers and forensic professionals. *Psychology Crime and Law*, *18*(2), 191–207. https://doi.org/10.1080/10683161003736744
13. Evett, I., Jackson, G., Lambert, J., & McCrossan, S. (2000). The impact of the principles of evidence interpretation on the structure and content of statements. *Science & Justice*, *40*(4), 233–239. https://doi.org/10.1016/s1355-0306(00)71993-9
14. Heavey, A. L., Turbett, G. R., Houck, M. M., & Lewis, S. W. (2021). Toward a common language for quality issues in forensic science. *Wiley Interdisciplinary Reviews Forensic Science*, *4*(4). https://doi.org/10.1002/wfs2.1452
15. Dunsin, D., Ghanem, M. C., Ouazzane, K., & Vassilev, V. (2024). A comprehensive analysis of the role of artificial intelligence and machine learning in modern digital forensics and incident response. *Forensic Science International Digital Investigation*, *48*, 301675. https://doi.org/10.1016/j.fsidi.2023.301675
16. Wilson-Wilde, L. M., Brandi, J., & Gutowski, S. J. (2011). The future of forensic science standards. *Forensic Science International. Genetics Supplement Series*, *3*(1), e333–e334. https://doi.org/10.1016/j.fsigss.2011.09.029
17. Ahmed Alaa El-Din, E. (2022). Artificial intelligence in forensic science: Invasion or revolution? *Egyptian Society of Clinical Toxicology Journal*, *10*(2), 20–32. https://doi.org/10.21608/esctj.2022.158178.1012
18. Casino, F., Dasaklis, T. K., Spathoulas, G. P., Anagnostopoulos, M., Ghosal, A., Borocz, I., Solanas, A., Conti, M., & Patsakis, C. (2022). Research trends, challenges, and emerging topics in digital forensics: A review of reviews. *IEEE Access*, *10*, 25464–25493. https://doi.org/10.1109/access.2022.3154059
19. Weinstein, J. B. (1969). The uniformity-conformity dilemma facing draftsmen of federal rules of evidence. *Columbia Law Review*, *69*(3), 353. https://doi.org/10.2307/1121012

20. Yadav, S., Yadav, S., Verma, P., Ojha, S., & Mishra, S. (2022). Artificial intelligence: An advanced evolution in forensic and criminal investigation. *Current Forensic Science, 1*. https://doi.org/10.2174/2666484401666220819111603
21. Kumar, M. (2024). An AI influenced social contract: An illusion of American constitution. *Social Informatics Journal, 3*(1), 31–37. https://doi.org/10.58898/sij.v3i1.31-37
22. Miller, T. (2018). Explanation in artificial intelligence: Insights from the social sciences. *Artificial Intelligence, 267*, 1–38. https://doi.org/10.1016/j.artint.2018.07.007. See further Sachoulidou, A. (2024). Harnessing AI for law enforcement: Solutions and boundaries from the forthcoming AI Act. *New Journal of European Criminal Law, 15*(2), 117–125. https://doi.org/10.1177/20322844241260114
23. Qi, A., Shao, G., & Zheng, W. (2018). Assessing China's cybersecurity law. *Computer Law & Security Review, 34*(6), 1342–1354. https://doi.org/10.1016/j.clsr.2018.08.007
24. Tynan, P. (2024). The integration and implications of artificial intelligence in forensic science. *Forensic Science Medicine and Pathology*. https://doi.org/10.1007/s12024-023-00772-6.
25. De Las Heras Ballell, T. R. (2019). Legal challenges of artificial intelligence: Modelling the disruptive features of emerging technologies and assessing their possible legal impact. *Uniform Law Review, 24*(2), 302–314. https://doi.org/10.1093/ulr/unz018
26. Mohsin, K. (2021). Artificial intelligence in forensic science. *SSRN Electronic Journal*. https://doi.org/10.2139/ssrn.3910244
27. Galante, N., Cotroneo, R., Furci, D., Lodetti, G., & Casali, M. B. (2022). Applications of artificial intelligence in forensic sciences: Current potential benefits, limitations and perspectives. *International Journal of Legal Medicine*, 137(2), 445–458. https://doi.org/10.1007/s00414-022-02928-5
28. Svantesson, D. (2021). The European Union artificial intelligence act: Potential implications for Australia. *Alternative Law Journal, 47*(1), 4–9. https://doi.org/10.1177/1037969x211052339
29. Paterson, M. *The Uses of AI in Government Decision Making: Identifying the Legal Gaps in Australia* (Vol. 89, pp. 647–665). https://mississippilawjournal.org/wp-content/uploads/2020/10/V89.4.3-PATERSON-8.15.20.pdf
30. Martin-Bariteau, F., & Scassa, T., eds. (2021, February 1). *Artificial Intelligence and the Law in Canada*. Toronto: LexisNexis Canada, SSRN. https://ssrn.com/abstract=3734675
31. Aditya, K., Grzonkowski, S., & Lekhac, N. A. (2018). Enabling trust in deep learning models: A digital forensics case study. *IEEE, 3*, 1250–1255. https://doi.org/10.1109/trustcom/bigdatase.2018.00172
32. Jo, W., Shin, Y., Kim, H., Yoo, D., Kim, D., Kang, C., Jin, J., Oh, J., Na, B., & Shon, T. (2019). Digital forensic practices and methodologies for AI speaker ecosystems. *Digital Investigation, 29*, S80–S93. https://doi.org/10.1016/j.diin.2019.04.013
33. Barik, K., Abirami, A., Konar, K., & Das, S. (2022). Research perspective on digital forensic tools and investigation process. In *Lecture Notes on Data Engineering and Communications Technologies* (pp. 71–95). https://doi.org/10.1007/978-3-030-93453-8_4
34. Costantini, S., De Gasperis, G., & Olivieri, R. (2019). Digital forensics and investigations meet artificial intelligence. *Annals of Mathematics and Artificial Intelligence, 86*(1–3), 193–229. https://doi.org/10.1007/s10472-019-09632-y
35. Salih, K., & Dabagh, N. (2023). Digital forensic tools: A literature review. *Journal of Education and Science, 32*(1), 109–124. https://doi.org/10.33899/edusj.2023.137420.1304
36. Parkinson, S., & Khan, S. (2024). The role of artificial intelligence in digital forensics: Case studies and future directions. *Assessment & Development Matters, 16*(1), 42–47. https://doi.org/10.53841/bpsadm.2024.16.1.42
37. Padma, K., & Don, K. (2022). Artificial neural network applications in analysis of forensic science. *Wiley Online Library*, 59–72. https://doi.org/10.1002/9781119795667.ch3

38. Mozaffari, A., Emami, M., & Fathi, A. (2018). A comprehensive investigation into the performance, robustness, scalability and convergence of chaos-enhanced evolutionary algorithms with boundary constraints. *Artificial Intelligence Review*, *52*(4), 2319–2380. https://doi.org/10.1007/s10462-018-9616-4
39. Walters, R., & Novak, M. (2021). Artificial intelligence and law. In *Springer eBooks* (pp. 39–69). https://doi.org/10.1007/978-981-16-1665-5_3
40. Novelli, C., Taddeo, M., & Floridi, L. (2023). Accountability in artificial intelligence: What it is and how it works. *AI & Society*, *39*(4), 1871–1882. https://doi.org/10.1007/s00146-023-01635-y
41. Zech, H. (2021). Liability for AI: public policy considerations. *ERA Forum*, *22*(1), 147–158. https://doi.org/10.1007/s12027-020-00648-0
42. Kingston, J. K. C. (2016). Artificial intelligence and legal liability. In *Springer eBooks* (pp. 269–279). https://doi.org/10.1007/978-3-319-47175-4_20
43. Gualdi, F., & Cordella, A. (2021). Artificial intelligence and decision-making: The question of accountability. *Proceedings of the . . . Annual Hawaii International Conference on System Sciences*. https://doi.org/10.24251/hicss.2021.281
44. Porayska-Pomsta, K., & Rajendran, G. (2019). Accountability in human and artificial intelligence decision-making as the basis for diversity and educational inclusion. In *Perspectives on Rethinking and Reforming Education* (pp. 39–59). https://doi.org/10.1007/978-981-13-8161-4_3
45. Smith, H. (2020). Clinical AI: Opacity, accountability, responsibility and liability. *AI & Society*, *36*(2), 535–545. https://doi.org/10.1007/s00146-020-01019-6
46. Rademacher, T. (2019). Artificial intelligence and law enforcement. In *Springer eBooks* (pp. 225–254). https://doi.org/10.1007/978-3-030-32361-5_10
47. Ponkin, I., & Redkina, A. (2018). Artificial intelligence from the point of view of law. *Rudn Journal of Law*, *22*(1), 91–109. https://doi.org/10.22363/2313-2337-2018-22-1-91-109
48. Butcher, J., & Beridze, I. (2019). What is the state of artificial intelligence governance globally? *The RUSI Journal*, *164*(5–6), 88–96. https://doi.org/10.1080/03071847.2019.1694260
49. Lima, G., Grgić-Hlača, N., Jeong, J. K., & Cha, M. (2022). The conflict between explainable and accountable decision-making algorithms. *2022 ACM Conference on Fairness, Accountability, and Transparency*. https://doi.org/10.1145/3531146.3534628
50. Shin, D., & Park, Y. J. (2019). Role of fairness, accountability, and transparency in algorithmic affordance. *Computers in Human Behavior*, *98*, 277–284. https://doi.org/10.1016/j.chb.2019.04.019
51. Wieringa, M. (2020). What to account for when accounting for algorithms. *ACM Digital Library*. https://doi.org/10.1145/3351095.3372833.
52. Lim, D. (2022). AI, equity, and the IP gap. *SMU Law Review*, *75*(4), 815. https://scholar.smu.edu/smulr/vol75/iss4/4 (Visited on 8 November 2024)
53. Bisoyi, A. (2022). Ownership, liability, patentability, and creativity issues in artificial intelligence. *Information Security Journal a Global Perspective*, *31*(4), 377–386. https://doi.org/10.1080/19393555.2022.2060879
54. Powles, J., Niklas, J., Sztandar-Sztanderska, K., Szymielewicz, K., Kubler, K., O'Reilly, T., & Kitchin, R. (2023). The algorithmic governance of administrative decision-making: Towards an integrated European framework for public accountability. In *The Algorithmic Governance of Administrative Decision-Making: Towards an Integrated European Framework for Public Accountability*. http://www.eurojus.it/pdf/III.2-GIULIA-SCHNEIDER-The-Algorithmic-Governance-of-Administrative-Decision-Making-Towards-an-Integrated-European-Framework-for-Public-Accountability.pdf
55. Doshi-Velez, F., Kortz, M., Budish, R., Bavitz, C., Gershman, S., O'Brien, D., Scott, K., Schieber, S., Waldo, J., Weinberger, D., Weller, A., & Wood, A. (2017). Accountability of AI under the law: The role of explanation. *arXiv (Cornell University)*. https://doi.org/10.48550/arxiv.1711.01134

56. Rosenbaum, H., & Fichman, P. (2019). Algorithmic accountability and digital justice: A critical assessment of technical and sociotechnical approaches. *Proceedings of the Association for Information Science and Technology*, *56*(1), 237–244. https://doi.org/10.1002/pra2.19.
57. Shah, H. (2018). Algorithmic accountability. *Philosophical Transactions of the Royal Society a Mathematical Physical and Engineering Sciences*, 376(2128), 20170362. https://doi.org/10.1098/rsta.2017.0362
58. Common Ethical Challenges in AI. (2023, September 29). *Human Rights and Biomedicine*. https://www.coe.int/en/web/bioethics/common-ethical-challenges-in-ai
59. Blodgett, M. S. (2011). Substantive ethics: Integrating law and ethics in corporate ethics programs. *Journal of Business Ethics*, *99*(S1), 39–48. https://doi.org/10.1007/s10551-011-1165-6
60. Gundugurti, P. R., Bhattacharyya, R., Kondepi, S., Chakraborty, K., & Mukherjee, A. (2022). Ethics and law. *Indian Journal of Psychiatry*, *64*(sup1), S7–S15. https://doi.org/10.4103/indianjpsychiatry.indianjpsychiatry_726_21
61. Hartman, L. (2000). Compliance versus integrity: The process of ethics integration. *Journal of Employment Discrimination Law*, 2(2), 157–159.
62. Madon, S., Burd, K. A., & Guyll, M. (2024). Do inconclusive forensic decisions disadvantage the innocent? *Law And Human Behavior*, *48*(1), 33–49. https://doi.org/10.1037/lhb0000553
63. Prins, C. (2018). Digital justice. *Computer Law & Security Review*, *34*(4), 920–923. https://doi.org/10.1016/j.clsr.2018.05.024
64. Moriarty, J. C. (2022). The inscrutability problem: From first-generation forensic science to neuroimaging evidence. *Duquesne Law Review*, *60*, 227.
65. Kroll, J. A. (2018). The fallacy of inscrutability. *Philosophical Transactions of the Royal Society a Mathematical Physical and Engineering Sciences*, *376*(2133), 20180084. https://doi.org/10.1098/rsta.2018.0084
66. Tsuda, T., Miyano, Y., & Yamamoto, E. (2023). Mistaken information can lead only to misguided conclusions and policies: A commentary regarding Schüz et al.'s response. *Environmental Health*, *22*(1). https://doi.org/10.1186/s12940-023-01013-7
67. Donoghue, J. (2017). The rise of digital justice: Courtroom technology, public participation and access to justice. *Modern Law Review*, *80*(6), 995–1025. https://doi.org/10.1111/1468-2230.12300
68. Rusakova, E. P., & Frolova, E. E. (2021). Current problems of digital justice in the BRICS countries. In *Smart Innovation, Systems and Technologies* (pp. 143–153). https://doi.org/10.1007/978-981-16-4621-8_12
69. Moliterno, J. E. (1996). On the future of integration between skills and ethics teaching: Clinical legal education in the year 2010. *Journal of Legal Education*, *46*(1), 67–78. http://www.jstor.org/stable/42893436
70. Curtis, R., & Christian, E. (2012). *Integrated Care: Applying Theory to Practice*. New York, NY: Routledge.
71. Kamila, M. K., & Jasrotia, S. S. (2023). Ethical issues in the development of artificial intelligence: Recognizing the risks. *International Journal of Ethics and Systems*. https://doi.org/10.1108/ijoes-05-2023-0107
72. Pavlova, T., Zarutska, E., Pavlov, R., & Kolomoichenko, O. (2019). Ethics and law in Kant's views: The principle of complementarity. *International Journal of Ethics and Systems*, *35*(4), 651–664. https://doi.org/10.1108/ijoes-04-2019-0080
73. Powles, G. (1999). Taking the plunge: Integrating legal ethics in Australia. *The Law Teacher*, *33*(3), 315–321. https://doi.org/10.1080/03069400.1999.9993039
74. Raj, M., & Seamans, R. (2019b). Primer on artificial intelligence and robotics. *Journal of Organization Design*, *8*(1). https://doi.org/10.1186/s41469-019-0050-0
75. Dignum, V. (2018). Ethics in artificial intelligence: Introduction to the special issue. *Ethics and Information Technology*, *20*(1), 1–3. https://doi.org/10.1007/s10676-018-9450-z

76. Rajasekar, K. P., & Vezhaventhan, D. (2024). Artificial intelligence revolutionizing legal and forensic practices: A comprehensive analysis. *IEEE, 3*(4), 1–12. https://doi.org/10.1109/icccnt61001.2024.10724685
77. Hunter, D., Bagaric, M., & Stobbs, N. (2019). A framework for the efficient and ethical use of artificial intelligence in the criminal justice system. *Florida State University Law Review, 47*, 749.
78. Kamran, S., Farhat, I., Mahboob, A. T., Kaur, A. G., Liton, D., Ghaffar, K. A., Rimsha, I., Irum, S., & Afifah, R. (2020). The impact of artificial intelligence and robotics on the future employment opportunities. *Trends in Computer Science and Information Technology*, 050–054. https://doi.org/10.17352/tcsit.000022.
79. Filipenko, N., Lukashevych, S., Andrieieva, O., & Ivanović, A. (2024). Application of artificial intelligence and information and communication technologies: Socio-ethical problems (review article). *Theory and Practice of Forensic Science and Criminalistics, 34*(1), 12–41. https://doi.org/10.32353/khrife.1.2024.02
80. Bogue, R. (2014). The role of artificial intelligence in robotics. *Industrial Robot the International Journal of Robotics Research and Application, 41*(2), 119–123. https://doi.org/10.1108/ir-01-2014-0300
81. Jeong, D. (2020). Artificial intelligence security threat, crime, and forensics: Taxonomy and open issues. *IEEE Access, 8*, 184560–184574. https://doi.org/10.1109/ACCESS.2020.3029280
82. Mir, U. B., Sharma, S., Kar, A. K., & Gupta, M. P. (2020). Critical success factors for integrating artificial intelligence and robotics. *Digital Policy Regulation and Governance, 22*(4), 307–331. https://doi.org/10.1108/dprg-03-2020-0032
83. Chen, Y., & De Luca, G. (2021). Technologies supporting artificial intelligence and robotics application development. *Journal of Artificial Intelligence and Technology, 1*(1), 1–8. https://doi.org/10.37965/jait.2020.0065
84. Iphofen, R., & Kritikos, M. (2019). Regulating artificial intelligence and robotics: Ethics by design in a digital society. *Contemporary Social Science, 16*(2), 170–184. https://doi.org/10.1080/21582041.2018.1563803
85. European Parliament Resolution of 20 October 2020 with Recommendations to the Commission on a Framework of Ethical Aspects of Artificial Intelligence, Robotics and Related Technologies, 2020/2012(INL). https://eur-lex.europa.eu/legal-content/EN/TXT/?uri=CELEX:52020IP0275 (Accessed on 14 November 2024)
86. Evett, I., Jackson, G., Lambert, J., & McCrossan, S. (2000). The impact of the principles of evidence interpretation on the structure and content of statements. *Science & Justice, 40*(4), 233–239. https://doi.org/10.1016/s1355-0306(00)71993-9
87. Angerschmid, A., Zhou, J., Theuermann, K., Chen, F., & Holzinger, A. (2022). Fairness and explanation in AI-informed decision making. *Machine Learning and Knowledge Extraction, 4*(2), 556–579. https://doi.org/10.3390/make4020026

12 Legal and Ethical Imperatives for Transparency

Devanshu Sharma[1] *and Dr. Navtika Singh Nautiyal*[2]
[1]National Forensic Sciences University,
Delhi Campus, New Delhi.
[2]Post Doctorate from MICHR, Italy.

12.1 INTRODUCTION

Transparency is the fundamental principle on which fast development in artificial intelligence occurs. In simpler terms, AI [1] refers to the extent to which systems can explain their methods, the process of reaching a conclusion, and, consequently, the outcome in words that are easily understandable and accessible. This is what explains how it works, its basis of decisions, and the degree to which such decisions meet societal, legal, and ethical requirements for the customer as well as the regulator.

12.1.1 Importance of Transparency in Fostering Trust

The wide adoption and usage of artificial intelligence technologies are actually based on the factor of trust. That means a black box where the outcome is achieved by the user without any ideas or reasons thereof. Systems that lack transparency raise concerns about biases and errors, which could lead to severe negative impacts, especially in fields such as healthcare, justice, and finance. Transparency gives users the means of seeing how the artificial intelligence algorithms work [2], allowing them to put greater confidence in the technology itself. For instance, consider the case where AI makes a decision on whether one is qualified to be granted a loan; that person may not understand how such a decision comes into being may ask why they are not eligible for the loan eligibility bias, among others. That shows that AI is transparent, and it explains why it chose credit-worthiness or income level, among which determines its verdict. Such endangerment ensures such assurance have trust from the user; trust and fairness are two reasons one gets to ensure ethical deployment.

DOI: 10.1201/9781003512820-12

12.1.2 Relevance of Legal and Ethical Considerations in Artificial Intelligence Systems

Legal and ethical frameworks have been put in place for the good use of AI technologies. In the nature of these structures, transparency is elemental. Fairness, accountability, and non-discrimination must therefore be present. Legal mandates, as reflected in provisions like the General Data Protection Regulation in the European Union, for instance, ensure that organizations provide reasons for any decisions that affect an individual's life made through the use of automated decision-making processes. Such provisions of law impinge on the benefits brought by transparency regarding the safety of basic rights, ensuring that an artificial intelligence system operates within its ethical considerations. This ethical issue ensures that artificial intelligence systems respect human dignity and rights, yet at the same time, supply information that could contradict or explain a decision that has already been made. For instance, if one computer recruitment tool refuses to recruit a person, it demands transparency so that the reason for rejection must be known and pursued – for example, if discriminatory practices have prevailed over the mind to reach the said decision. This prevents the misuse of AI technologies.

12.1.3 Transparency's Role in Digital Forensics and Decision-Making Systems

In a digital forensic approach, this means that the evidence that AI generates must be transparent enough to rely and hold for the court. AI solutions increasingly employ analyses and identification of cyber threats using digital artifacts while, at the same time, tracing criminal activities. The methods and procedures followed in these AI environments need to, therefore, be traceable; clear documentation for court and prosecution purposes is important to discern from them.

In addition, transparent decision-making systems allow accountability because parties with a stake in the outcome can check the outcome and appeal errors or biases. To give an illustration, in criminal justice, AI can predict recidivism rates or determine guidelines for sentencing. A transparent system ensures that such decisions are not made on flawed or biased data but on objective grounds, which enhances not only the integrity of the justice system but also its protections against potential misuse of AI.

12.2 LEGAL FRAMEWORKS FOR ARTIFICIAL INTELLIGENCE TRANSPARENCY

The more advanced the AI, the greater the need for a legal framework in AI transparency. Every government, coupled with regulatory bodies, drafts and implements legal frameworks that ensure that problems associated with concerns over opacity, bias, and accountability are addressed by the AI system. These frameworks ensure that AI technologies are developed in a fair, explainable, and aligned manner with accepted ethical norms and societal expectations. This chapter will discuss the legal frameworks, examine international AI regulations, and add case studies demonstrating the impact of a lack of transparency in AI. Critical Legal Frameworks on the Mandate of AI Transparency: Several notable legal frameworks give special emphasis to transparency as one of the standards applied to AI systems. For instance, the General Data Protection Regulation of the European Union and the California Consumer Privacy Act of the United States can be cited as significant regulations.

12.2.1 Global Laws and Regulations Promoting Transparency in Artificial Intelligence

12.2.1.1 General Data Protection Regulation (GDPR)

The GDPR, adopted by vote in 2018, has set an international standard for the protection of data and privacy. It includes regulations that directly impact artificial intelligence systems [3], especially when related to automated decision-making processes. Article 22 grants data subjects the right to object to being subjected to decisions based exclusively on automated processing, including profiling, where the information significantly affects them. Furthermore, what follows is a call for transparency:

A) **Right to Explanation:** This can be concluded from the reading of Articles 13–15, which establish that individuals have the right to an explanation when automated decision-making results from how the application works and affects them.
B) **Accountability and Traceability:** Organizations must demonstrate observance of data protection principles in GDPR by appropriate and accurate data processing records. These regulations provide assurances that AI/deep learning systems functioning in the European Union will provide subjects with significant knowledge of the reasons and information that comprise these operations, as well as how these algorithms will utilize the user's data.
C) **Data Protection Impact Assessments (DPIAs):** This covers the impact of automated processing involved in high-risk AI applications, including profiling on individual rights.

12.2.1.2 California Consumer Privacy Act (CCPA)

CCPA protection of consumers' rights in personal data within California [4] started to apply in 2020. It also places a mandate on organizations to inform about the collection and use of data and where such data is shared with third parties. Most of its transparency-related requirements indirectly apply to AI processing consumer data, not strictly targeting AI. Key provisions are listed below:

A) **Data Access Rights:** Consumers have the right to request access to categories and specific pieces of personal information collected.
B) **Disclosure Duties:** Businesses have the duty to let consumers know every possible purpose for the information collected, including AI-based decision logic. The CCPA becomes the base in the United States that in itself drives other states to come up with similar acts and also quickens the transparency process for AI.

12.2.2 Other International Artificial Intelligence Laws on Transparency

Aside from GDPR and CCPA, many countries, along with international bodies, are developing AI-specific laws and regulations based on the principles of fairness, accountability, and transparency.

12.2.2.1 EU AI Act

The European Union AI Act is the first comprehensive regulatory framework for AI systems. Introduced in 2021, AI applications are broken into four broad risk classes: **minimal, limited, high, and unacceptable**. In high-risk AI systems like healthcare, law enforcement, and financial services, the Act includes the following:

A) **Transparency Obligations:** The developer or provider of the AI system must reasonably provide documentation detailing how the AI system functions and makes its decisions.
B) **Human Control:** AI systems shall be created to allow for easy human control/intervention to override.
C) **Accountability Controls:** Organizations shall ensure that the AI systems are of high ethical and legal standards throughout all lifecycle stages.

The Bill is expected to set an international benchmark for AI technology regulation.

12.2.2.2 Other Emerging International Efforts

A) **Singapore:** In its Model of AI Governance Framework, the country focuses on transparency through explainability as well as stakeholder communication. It balances innovation with ethical concerns, promoting voluntary guidelines that encourage organizations to adopt best practices for transparency.
B) **India:** The legislation specific to AI remains at a very early stage of regulation in India. However, the draft Digital Personal Data Protection Bill throws open several principles that can guide regulation on AI; it talks about data transparency and accountability.
C) **United States:** Besides state-specific efforts such as CCPA, federal efforts include the Algorithmic Accountability Act, which has provisions for duties of transparency in high-impact AI systems. The lack of federal law to govern AI directly causes the practice to vary in its enforcement despite state-by-state progress, thereby creating a fragmented [4] approach.

12.2.3 Case Studies: Consequences of Non-Transparency in Artificial Intelligence

I. **COMPAS and Algorithmic Bias**
This algorithm in COMPAS [5], however, when used in the US sphere of criminal justice, was met with much criticism and opposition from legal entities. The mechanism about it not being transparent enough evoked the possibility of racial bias in parole decisions. Moreover, the incoherent

explanation of the predictions by COMPAS has resulted in a deficiency of trust in the system and put emphasis on the aspect of transparency in urgent applications with high stakes involved.

II. **Facebook and GDPR Fines**
GDPR is pushing against the algorithmic operations of Facebook [6], and others, designed for targeted advertisement. In 2021, the Irish Data Protection Commission issued an order requiring Meta to pay a €265 million fine after it was alleged to have failed to be transparent about how users' data was used in targeted advertisements. Such a case evidences transparency in disclosure with AI-based operations.

III. **Clearview AI and Privacy Concerns**
Clearview AI, one of the firms offering facial recognition, has faced legal issues in different jurisdictive systems owing to their lack of transparency. The organization's failure to inform people about the collection and use of their biometric data led to cases in the European Union and in the United States. Regulatory bodies such as the French CNIL imposed sanctions, emphasizing the need for transparency in artificial intelligence systems that handle sensitive information.

12.2.4 Comparative Analysis of Legal Approaches

Overview of the different countries' legal systems and development, as well as enforcement levels on the regulation of AI transparency:

I. **European Union:** The EU sets the gold standard for AI governance with its GDPR, EU AI Act, and DSA. Effective, enforceable laws like GDPR and the soon-to-be-out AI Act mark a landmark in such subjects of transparency and responsibility. All these approaches make the EU a global leader in ethical AI practices.

II. **United States:** The lack of federal law to govern AI directly causes the practice to vary in its enforcement despite state-by-state progress. This makes the United States' approach a fragmented one.

III. **India:** India remains at the very earliest stages of AI regulation, but it is open to massive transformation as it solidifies in the Digital Personal Data Protection Bill. This inequality necessitates standardized global norms to ensure the same application of the principle of transparency across borders.

12.3 ETHICAL FOUNDATIONS OF TRANSPARENCY

Most of those challenges have an anti-thetical variant in the emerging call for more transparency in AI: an ethical instead of legal or technical challenge. That is to say, there will be correspondence between societal values and the nature of the AI systems. Human dignity will be respected and trust built. This chapter critically explains why frameworks of ethics stand supportive of frank and morally needed explanations in AI but also, in contrast, reveals some of the problems when such explanations raise problems with proprietary or security considerations.

12.3.1 Ethical Theories for Transparency

Ethical theories stand as a rational way to understand why transparency is necessary in AI systems [7]. The two leading theories – consequentialism and deontology – strike a balance with the contrasts of their respective views on the very need for transparency.

I. *Consequentialism*
   It's a measure of the morality of an action based on its result. Considering this theoretical framework, transparency will be morally justified, as it shows several beneficial effects: fairness, accountability, and trust. Thus, for example, once the AI system is transparent, users and stakeholders have more information and knowledge about decisions being made; therefore, possible negative consequences resulting from errors or bias in AI systems are minimized. Finally, transparent AI systems will make the stakeholders aware and redress inequalities so that goods will be served for society.
   This captures consequentialist [8] reasoning within the context of healthcare AI systems; if diagnostic tools are transparent, enabling them to explain a recommendation, better patient
   outcomes are produced by enabling informed consent and collaborative choice. Opaque systems risk harming through misdiagnosis or the continuation of health inequities.
II. *Deontology*
   Deontological ethics are concerned with duties and rules: it is a doctrine that conceives actions to be intrinsically possessed of a moral character of right or wrong, quite independently of their consequences. Transparency, in the deontological approach, turns out to be a duty based on respect for autonomous agency and the principle of justice. This position asserts that people have a right to know the steps behind decision-making affecting them, regardless of the concrete benefits that may accrue from such disclosure.

For example, deontological ethics are applicable in the administration of justice while AI systems are expected to explain all analyses of forensic evidence because secrecy defeats the balance and truthfulness that justice mechanisms ought to maintain.

12.3.2 Duty to Explain the Results of AI Analysis

This integration of AI into the strategic sectors, including finance, healthcare, and public administration, brings with it the moral obligation to explain the reason behind this since decisions that are made by AI have far-reaching impacts as they may dictate destinies even in lending loans and medical diagnosis and criminal sentences.

12.3.3 Human Autonomy Evolution

Respect for individual autonomy requires that individuals be informed so that they can make decisions in their lives. Any system of AI that becomes influential in deciding, whether it is on what kind of welfare programs to choose or what kind of treatment is then required, transparency makes sure that people do not function simply as receivers of results but are instead actual participants in decision-making processes.

12.3.4 Improving Organizational Accountability

Transparency is actually the fundamental component of the accountability framework concerning organizations. Organizations of the AI-based [9] type have to explain their thought process before regulatory bodies, other stakeholders, and the remaining population. At such crucial moments when the systems are not transparent, they would be vulnerable points for allegations against them of bias, negligence, or even malpractice, especially in areas like online recruitment or the distribution of resources during crises.

In certain sensitive domains, such as healthcare and law enforcement, these ethical obligations to transparency in AI decision-making are crucially important. The absence of such transparency can lead to the erosion of trust between patients and healthcare providers and may also lead to unjust consequences for already vulnerable populations affected by the algorithms used in risk assessments or surveillance for law enforcement.

12.3.5 Balancing Ethical Considerations with Conflicting Interests Review

While transparency could be an ethically necessary requirement, it often conflicts with other legitimate interests in protecting rights of intellectual property, national security [10], or even the avoidance of misuse. Such conflicts raise very difficult ethical dilemmas to be weighed judiciously.

Many AI systems are proprietary. Algorithms may represent intellectual property. Companies may not want full openness as such systems provide a competitive edge. However, if unethical implications arise from such opaque systems that affect public welfare, they must be addressed and allowed to be corrected. Credit scoring algorithms are often proprietary and have been criticized for the lack of transparency in decisions about one's financial life that affect them drastically.

Defense and surveillance or counter-terrorism AI systems are classified, with their workings kept secret for national security reasons. While it may be justifiable to keep this secret, it throws open many ethical dilemmas—most notably those concerning liability and exploitation of the systems. For instance, public facial recognition in the framework of surveillance should be open, but not revealing about how it works.

12.3.6 Risk of Misuse

It risks laying its algorithms open to reverse-engineering for calamitous uses-for example, consider a fraud detection system that publishes the logic by which the system makes its decisions totally, and then malicious actors take advantage of its weaknesses.

12.3.7 Balancing Ethical Trade-Offs

To alleviate these dilemmas, stakeholders may consider strategies to balance transparency with countervailing interests.

- **Tiered Transparency**: Elaboration to regulators, summarization to other stakeholders in the industry, and public accounts accessible to all.
- **Independent Third-Party Audits**: Independent audits can provide the means of transparency for secret or proprietary algorithms without infringing on intellectual property rights or security issues.
- **Regulatory Oversight**: National governments and international bodies may enact legislation mechanisms that could enable enough extraction of transparency without infringing on intellectual property rights or security issues.

12.4 CHALLENGES TO ACHIEVING TRANSPARENCY

It's a fundamental foundation on which trust and accountability in AI are constructed. However, it has been a pretty tough challenge to put into practice. Technical, corporate, and systemic ones intertwine in ways that make it hard to render clarity in AI systems. This chapter discusses a couple of the intrinsic issues related to AI technologies, proprietary algorithm fears, and the absence of uniform global standards.

12.4.1 Technical Challenges

I. *Black-Box AI Models*
AI systems [11], especially those designed using deep learning techniques, are mostly black-box entities as they seem opaque and lack meaningful interpretability. Even though they perform very well in areas of pattern discovery and decision-making, it's certain they lack intuitive interpretability, thus greatly challenging the issue of transparency.

II. *Delicacies of Neural Networks*
Neural networks are at the heart of most of the currently successful artificial intelligence systems because they process information layer by layer within interconnected nodes. Although this architecture gives spectacular accuracy, it develops intricate decision paths simultaneously, which are very hard to trace or explain. For instance, an artificial intelligence program used to diagnose medical [12] conditions may consider millions of evidence pieces before making a particular forecast, making it pretty impossible to decipher the reasoning behind a particular decision.

III. *Lack of Explainability in Machine Learning*
Unlike rule-governed traditional software systems, machine learning-based models [13] learn from large datasets and develop patterns without explicit human interference. This autonomy leads to a disconnection between the mechanisms by which AI systems operate and the ability to explain the decisions they make. Sometimes, domain experts are unable to fully understand the reasoning behind the generated output from AI, particularly if the scenario is built using a convolutional or recurrent neural network.

IV. *Innovative Solutions and Their Limitations*
Despite these approaches, for example (Shapley additive explanations (SHAP [14]) and local interpretable model-agnostic explanations (LIME), they that strive to bridge the gap, have certain limitations. These methods provide approximations rather than actual explanations and rely on specific expertise for understanding them. Besides, the XAI techniques can result in computational overheads, thus not being suited for use in real time.

12.4.2 Resistances from Organizations towards Opening Up of Their Secret Algorithms

I. *Intellectual Property Considerations*
Artificial intelligence algorithms typically have intellectual properties for businesses, and this is what distinguishes them from competitors. For example, the algorithms a large technological firm uses to deliver personalized recommendations or block fraud are closely guarded secrets. Any information about such systems can be disclosed with the intent that their competitors can replicate or top them, thus threatening their market dominance.
II. *Potential Vulnerability*
Another reason for corporate reticence is the potential reality that openness may highlight weaknesses. Fully exposing an algorithm for AI-driven credit scoring could give criminals [15] access to manipulating their credit records such that they get positive outcomes. The inner functioning of a cybersecurity AI might be used by a hacker as knowledge that can help bypass its securities.
III. *Regulatory Resistance*
In fact, even while private enterprises resist it, regulators and other stakeholders are increasingly demanding transparency. Even while the European Union AI Act [16] looks forward to making explainability compulsory for particularly high-risk AI systems, tension betrays corporate interests and societal demands for accountability.

12.4.3 Lack of Standardization and Global Agreement on Transparency Norms

Lack of standardization and global agreement on transparency norms and without widely understood standards on transparency, the burden is exacerbated for AI developers and regulators. As "different regions and industries have different approaches towards transparency," discrepancies and ambiguity abound.

I. *Patchwork Legal Systems*
The landscape of regulation globally varies significantly and becomes highly heterogeneous. In the European region, it leans more towards transparency, as it outlines the General Data Protection Regulation and the proposed AI Act; whereas in the United States, it continues to remain sectoral, and the market is driven by sector-specific regulations. India too still stands at the beginning of comprehensive AI policy building. Such heterogeneity creates a maze of different rules for global firms to cut through.

II. *Lack of Technical Standards*
Then, there is the technical standardization as far as the aspect of transparency is concerned. ISO and IEEE are, however, working on standards for AI. Progress has been slow. In the absence of generally accepted guidelines, it cannot be ascertained how developers can apply transparency so that the various stakeholders are satisfied.

III. *Cultural and Ethical Differences*
Additional factors that make coordination at the global level difficult are cultural and ethical values related to transparency. For example, rights of individuals and their self-autonomy form the central theme in Western perspectives towards transparency, but collective welfare and national interests are centered at Asian levels. Such diversity in values places severe impacts on the regulatory priorities as well and thus provides further barriers to harmonization.

12.4.4 To Overcome the Challenges

I. *Advancement of Technical Solutions*
Develop mature forms of XAI techniques by explaining complex models in a way that does not degrade performance. Efforts should also be made so that user-friendly tools are developed through which any non-expert person can understand what AI decisions are taking place behind.

II. *Balancing Transparency with Secrecy*
Organizations can utilize third-party audits as one method of expressing openness standards, yet at the same time keep the selective information secret. Likewise, differential transparency can obtain balance by calibrating the degree of disclosure according to the needs and demands of different stakeholders.

III. *Global Standards*
Industry leaders should coordinate with policymakers in the development of global standards. Organizations like the OECD and the United Nations happen to be best suited for this purpose.

IV. *Education and Awareness*
Targeting all stakeholders ranging from developers to legislators and wider populations, education and awareness about the importance of AI transparency will be emphasized. More knowledge can create demand for more transparent systems and have the strength and influence the necessary control and make organizations accountable.

12.5 IMPLICATIONS OF LACK OF TRANSPARENCY

Such a system of artificial intelligence represents a lack of transparency in its substantial impacts on ethical, social, and legal dimensions. As these systems play significant roles in sectors such as healthcare, finance, and law enforcement for example, where they generate decisions, these opaque systems have to be addressed immediately.

These risks include algorithmic bias related to the process of making decisions, discriminatory practices targeted at some people, loss of trust, and various liabilities that could be acquired in court. In society, artificial intelligence's full scope would be impossible without transparency. Its inconsistency is manifold and prevalent in many fields.

12.5.1 Risks of Algorithmic Bias, Unfair Decision-Making, and Discrimination

Black box AI systems are opaque. This is to say, one cannot gain significant insight into what went into the decision. Often this opacity hides bias within an algorithm [17] should its outputs be adverse for some groups. Bias can be traced back to the training data, possibly mirroring historical injustices or poor algorithmic designs that amplify such inequalities. Another criticism that AI-based recruitment algorithms have is accentuating biases in gender and race. These systems denied opportunities to specific candidates based on the resultant patterns generated by the algorithm that systematically pushed out better-qualified candidates from under-represented groups.

The most heavily processed features concerning decisions, for instance, on issues of health or the criminal justice, have terrible consequences that are largely adverse. Inaudible AI diagnostic systems could diagnose wrong and/or offer access to unfair treatment. Predictive policing algorithms have been faulted for continuing systemic bias issues in the algorithms that their use in discriminatory enforcement practices is likelier.

Similarly, a lack of accountability destroys transparency. When the process through which these decisions are made is not aired in the most transparent way possible, it becomes difficult to audit or reverse adverse effects. In this circumstance, the damage that an unfair or prejudiced AI system might inflict can go deepened, further reducing public trust.

12.5.2 Loss of Trust among Users and Stakeholders

Building the basis of trust through openness is important, the absence of which can cause serious damage to confidence in AI systems. Users would not like to communicate and interface with technologies perceived as opaque or untrustworthy. For instance, a customer might turn down an AI-based credit scoring system because they do not have an idea of how their credit-worthiness is being measured. So, regulators, policymakers, and investors may not invest in organizations that cannot provide insights into what is happening behind the AI activities [18].

Lacking trust, adoption and innovation can also be slowed. Organizations have no control over AI-driven decisions that have the potential for reputational damage and decreased market competitiveness. Publicized examples of failed opaque AI, such as undesirable medical diagnoses or wrongful arrests due to misapplied algorithms, amplify the liabilities associated with reputational dangers in the absence of transparency. Such examples cause skepticism not only regarding certain systems but even more general AI technologies.

12.5.3 Legal and Ethical Consequences of the Black Box

There are significant challenges on both legal and ethical fronts. Legally, many jurisdictions require organizations to explain automated decisions that affect a person. For instance, the General Data Protection Regulation in the European Union dictates the "right to explanation," making it mandatory for companies to reveal the reasoning behind automated decisions. The absence of such explanations may lead to a heavy price to pay in terms of fines and legal liabilities.

Ethical implications are equally profound. Black-box AI violates norms of fairness, accountability, and respect for human agency. In contexts in which decisions made by AI potentially implicate rights to life, liberty, and security of person –through denial of loans, job opportunities, or parole – opacity violates ethical requirements for just and fair decision-making. For example, in criminal justice systems, the use of opaque risk assessment tools to decide sentences is profoundly problematic along lines of transparency and due process.

Real-world case studies illustrate such results. One notable case was the US criminal justice's proprietary AI system that assigned risk scores to defendants. This system, which, having been documented to contain racial biases, provided fertile ground for valid criticisms and lawsuits, which point nicely to the dangers of opaque AI in high-stakes contexts.

12.5.4 Implication of the Black Box

On the other hand, the full effects of such shadowy AI systems turn out to be such that transparency is urgently called for as the guiding principle in development and deployment. Such a call would be a comprehensive multidimensional approach combining technical, organizational, and regulatory measures. Techniques like XAI can unravel complex models, while regulatory frameworks encourage organizations to make transparency a priority.

Accountability: The mechanisms may include audits, impact assessments, and independent oversight. It is through embedding transparency into the design and governance of AI systems that organizations will better ensure risks in terms of bias, unfairness, and discrimination when the use of the same is fostered.

12.6 PROPOSED GUIDELINES AND STANDARDS FOR TRANSPARENCY

This means that the transparency of AI systems is both legally possible and ethically secured. The penetration of AI in various sectors without definite parameters and standards will pose severe social, economic, and legal challenges for some sectors. To that end, this chapter assesses the international framework to date, further suggesting clearer operational transparency, based on mutual cooperation among governments, industry players, and universities in setting adequate standards.

12.6.1 Analysis of International Rules

International efforts have given greater importance to transparency as the imperative to promote the ethical development of artificial intelligence. In this connection, it is highly important that 46 countries have adopted OECD AI Principles, focusing public attention on both transparency and explainability through AI systems that are intelligible to users and stakeholders. OECD guidelines elaborate further on the necessity to provide adequate information regarding the objectives, workings, and impacts of AI systems so that their use can be aligned with societal values.

The EU has taken a leadership role in shaping the development of AI regulation through the AI Act: an incredibly comprehensive and dense regulatory framework towards the achievement of trust and accountability for AI systems. For instance, the AI Act-based grouping of AI applications is based on the risks they generate, with strict transparency requirements such as law enforcement, healthcare, and financial services. For example, vendors operating in these areas are compelled to reveal in full detail their algorithms, the training sets of data, and the decision-making processes.

Furthermore, the General Data Protection Regulation imposes the "**right to explanation,**" thereby making the organization liable for the provision of clear information on the automated decision-making processes. A right to explanation provides that decisions based upon AI that affect rights and freedoms should be transparent and accountable.

The United States is more sector-specific, and different sectors have variations with regard to their regulation. For example, the Federal Trade Commission attaches more importance to transparency in artificial intelligence because it has prevented any deceptive practices, especially those targeting consumers. However, failure to fully tackle transparency issues results from this lack of an integrated federal framework.

India and other emerging economies are now starting to outline their AI policy, beginning to appreciate the importance of transparency. The Draft National Strategy on AI in India promotes the use of transparent AI systems to create trust and accountability. Some of these regulatory measures are still at a nascent stage of development, and therefore, there is an urgent need to build concrete oversight.

12.6.2 Recommendations for Implementation of Transparency in Artificial Intelligence Systems

Such a multidimensional objective of achieving transparency in AI systems involves technical, organizational, and policy-driven strategies [19]. Some of the recommendations made for effective operationalization of transparency are below:

- **Use Explainable AI (XAI) Techniques:** Explainable models like Shapley additive explanations (SHAP), local interpretable model-agnostic explanations (LIME), or decision trees enable demystifying complex algorithms. This tool will allow the developers to explain through various mechanisms how inputs affect outputs, thus making AI systems accessible to non-expert stakeholders.

- **Embed Transparency into the Design:** Transparency cannot be a supply-and-demand add-in, but integrated into the AI development cycle. Developers need to explain in nontechnical terms the thought process behind the code, through interpretable algorithms, and maintain comprehensive user guides so that end users understand how the system works.
- **Create Audit Mechanisms:** Independent body assessments may also verify whether the requirements of openness are met. Such assessments ought to take into account not only the technologies of AI systems but also their social and ethical implications so that they truly comply with the set recommendations.
- **Formulate Data Transparency Requirements:** Institutes have to disclose the origin, quality, and representativeness of data used to train AI systems. Disclosure of such information will help curb discrimination and bias; hence, a relief is taken as it helps bring accountability.
- **Accessible User Interfaces should be designed with end users in mind so that they can have useful interactions with AI.** With such interfaces being user-friendly, and after clear reasoning for any decisions made, they inspire trust within the system.
- **Include Mechanisms for Feedback:** Stabilized feedback systems will help users determine anomalies or biases in systems and open up opportunities to make iterative improvements and raise the bar on transparency.

12.6.3 Government-Industry-Academia Collaboration

Internationally accepted standards for artificial intelligence transparency can only be developed through a collaborative approach involving governmental organizations, industry players, and educational institutions. Each stakeholder brings unique knowledge and resources to the table, opening possibilities for collaborative benefits.

I. *Governmental Role:*
Policymakers should spearhead international harmonization efforts. They should establish transparent rules and regulations, provide incentives for compliance, and encourage international cooperation. Multilateral organizations such as the United Nations and the OECD can play [20] a role in facilitating dialogue and bringing countries to a consensual agreement.

II. *Industry Involvement:*
Transparency has to be encouraged on the part of the private sector through the development of AI models by organizations, enactment of best practices, and sharing knowledge in industry coalitions. Industry leaders can set a good example as they can vow to adhere to ethical AI standards with clear principles set towards transparency, publicly shown. It lays down the conceptual and technological preconditions for transparency to be made possible. Research groups can now produce new solutions in explainable AI and can evaluate the social impact of transparency initiatives while teaching the next generation of AI professionals how to focus on transparency within their work.

III. *Public-Private Partnerships*
Acceleration of progress can be hastened through cooperative work among government, industry, and academia. Examples of cooperation include cooperative research projects, standardized testing frameworks, and shared datasets that better advance common understanding of transparency challenges and solutions.

12.6.4 Towards a Unified Standard

Of course, in some areas, there are positive tendencies, but so far there is no global standard for AI transparency. Different legal structures and cultural approaches present significant barriers to the development of international norms. Still, the growing sensitivity to the problem of transparency becomes an important factor stimulating unification.

Such global frameworks may have their standardization templates, such as OECD AI Principles. With such an alignment with domestic policies, a country can thus set up a harmonious regulatory framework that will support transparency while allowing for innovation.

12.7 FUTURE TRENDS IN AI TRANSPARENCY

The history of the development of AI technologies has solidly consolidated transparency among the big four principles of ethical, responsible AI development. Innovations that not only erase current practices of transparency but also open new doors for further developments pose problems. Thus, this chapter discusses leading-edge technology, the future of standardization on transparency, challenges and opportunities in the creation of norms accepted worldwide, and other issues.

12.7.1 Emerging Technologies to Enhance Transparency

Another approach very well known to achieve greater transparency in AI is through blockchain technology. Its primary role is as an immutable and decentralized ledger that tracks the entire lifecycle of all AI systems. It allows recording in an auditable yet non-repudiable way everything from data acquisition and preprocessing stages through the training and deployment of the model.

In this connection, blockchain technology can meet many concerns about data origin by tracing and ethically acquiring the training datasets. Besides that, smart contracts – the self-executing algorithms operating on blockchain networks – can help in the ease of compliance checks to ascertain whether the AI system meets developed transparency requirements before its deployment. It is an approach that promotes accountability but also eases the burden of administration for the organization.

Notable among these tendencies is the presence of explainable AI methods. Such complex methods as SHAP and counterfactual explanations are increasingly necessary to explain complicated algorithms. Besides increasing the interpretability capabilities of the AI model, these tools enable developers to detect biases and correct them for AI systems that do not violate ethical rules.

However, on the other end of the scale, federated learning and differential privacy – methods that focus on data privacy – cater to the tension between the need for a notion of transparency versus the protection of data. They permit models to learn from decentralized datasets without undermining individual confidentiality as they preserve transparency and keep confidentiality among users.

12.7.2 Hypothesized Future Legal and Ethical Frameworks

This future trajectory of AI transparency will shift both the legal conditions and the ethical ones. Current frameworks – for example, the European Union's AI Act and the General Data Protection Regulation – provide a starting point for broader, more specific requirements in terms of greater obligations by way of more concrete transparency. It is likely that such regulations will fill out in scope and detail as AI invades additional realms of society.

A visioned evolution will include worldwide harmonization of AI regulation. With international institutions already asking for "**harmonized principles**," even countries could harmonize on a common set of standards towards transparency: at least meaningful interpretability requirements on models, full records of all decision-making operations, and feedback systems for the users.

In the ethical dimension, it is expected that transparency will be imbued more visibly in the values of society. For instance, in healthcare AI, the principle of autonomy should be stood on its head; in financial AI, equitable access will be the core. This injects sector-specific nuances and calls for flexible yet comprehensive guidelines to deal with various considerations.

12.7.3 Issues while Formulating Consensus-Based Norms

Although there is a fair consensus on the need for transparency, attaining commonly accepted standards proves difficult. First, there is no universal agreement on what constitutes sufficient transparency. The cultural, economic, and legal diversity among nations prevents uniformity: for instance, the industry-centric model pursued by the United States is a dramatic contrast to the regulation-focused structure adopted by the European Union—a structure comprised of differences that manifest on the ground, say, with regard to various measures for achieving transparency.

There is also a lot of technical complexity involved in much artificial intelligence. For example, deep learning models, especially those running neural networks, are inherently "black boxes." This means that the decisions reached are typically opaque to the designers of the system. More generally, opening up technical details to non-experts is quite a significant challenge further down this line.

Yet, on the other hand, proprietary concerns are involved. Parties regard their own algorithms as confidential and shun open access lest they portray the vulnerability of security in their system's architecture. A combination of openness and protection of intellectual property will require creative methods such as selective disclosures or peer review but in the form of confidentiality agreements.

12.7.4 Area for Improvement

Despite the challenges, there are still many fields yet to be investigated before the principles of transparency can be applied more widely. Cooperation between governmental bodies, universities, and the private sector is promising. Collaborations and sharing resources may raise highly sophisticated frameworks that are both realistic and feasible to apply. For example, public-private partnerships can finance research on explainable AI approaches or set up regulatory sandboxes focused on transparency.

Another area of interest relates to technological innovation. The use of synthetic data in conjunction with automated compliance systems may be powerful levers to enhance the enforcement of standards of transparency. Advancements in natural language processing will also ultimately be able to make AI-generated explications of decision-making algorithms understandable to non-expert stakeholders, thus raising trust among them. Public engagement is actually one of the most important channels for progress: informing citizens about the importance of transparency allows government officials to cultivate grassroots advocacy for regulations, and citizen contributions can make the standard-setting process more responsive to popular needs and values.

12.8 CONCLUSION

Artificial Intelligence translucency has now emerged as the principle that ensures that promises of transformative technologies will also be in line with law, ethics, and societal expectations. Building trust with users and protection from algorithmic bias and discrimination requires artificial intelligence systems to be transparent and interpretable. Artificial intelligence impacts many sectors, including healthcare, finance, forensics, and governance – making accountability, equity, and compliance with regulatory standards highly dependent on it.

12.8.1 Meaning of Transparency Legal and Ethical Issues

In legal terms, transparency in AI is evidence of the achievement of compliance. The thrust of guidelines like GDPR in Europe and CCPA in California, or newly emerging legislation around the EU AI Act, emphasizes the importance of documentation around the workings of AI systems and that their rights are taken care of. Moreover, the legal requirements related to traceability and reproducibility also favor introducing transparency into AI applications, particularly where such decisions significantly affect people or communities. Lack of transparency has already led to some egregious cases and monetary sanctions so that transparency becomes an important aspect in the acceptance of ethical AI systems.

Ethically, standards mean transparency and equity – responsibility and respect for individual autonomy at basic levels.

The more transparent the framework, the easier it will be to explain why a decision was taken and minimize bias to ensure that results follow the norms of society and ethics. Healthcare and criminal justice form the core fields where transparency in AI decision-making reflects a strong need for ethics: opaque algorithms may have dramatic impacts on people's lives and socioeconomic conditions.

12.8.2 Barriers to Transparency End

Perhaps achieving transparency in AI systems entails, in itself, a great challenge. Such an achievement of transparency is technically hindered as the problems involved are complex because most complex machine learning architectures are "black boxes." Despite the, strong capabilities such models possess, they lack the inherent interpretability necessary for transparency; thus, additional instruments, such as XAI methods, are required for the same purpose. In fact, the challenge that influences initiatives to improve transparency is specifically an aspect of copyright protection and the reluctance of corporations to disclose proprietary algorithms. For most organizations, their algorithms represent key trade secrets, with the fear of losing a certain advantage should they be released. A balance between the need for transparency and intellectual property rights necessitates innovative approaches like third-party evaluations or regulated disclosures under nondisclosure agreements.

One of the key challenges for this question is that there does not exist a universally agreed minimum level of transparency. Policies vary with national and regional differences in cultural, legal, and economic contexts. This leads to much inconsistency. It becomes hard to establish clear transparent practices, especially in the context of an increasingly integrated world economy.

12.8.3 Key Recommendations

The project will be from different sectors, and the practice needs to involve government agencies, corporate heads, and learning institutions defining the right practices that enhance transparency. Public-private collaborations help in the development of explainable artificial intelligence methodologies. Regulatory sandboxes can be a safe space to test and improve practices on transparency. Organizations such as OECD and UNESCO [21] can then lead in harmonizing global standards while taking regional differences into account so that one has consistency without unnecessary complexity. Equal investment in education and awareness is essential. Inform policymakers and developers on how to put in place openness effectively. In parallel, public involvement must take place so that the transparency measures are based on the expectations of society for their achievement.

12.8.4 A Call to Involvement

It is a long and uphill journey to achieve transparent AI systems, but the outcome is grand. Policymakers should therefore be pressured to create requirements for detailed and enforceable regulations requiring complete transparency at every step of AI development and implementation. Regulation must be implemented as the mere minimum that is required, whereas transparency must become the basis of ethical innovation for developers. That end must be furthered while research in explainable AI and related topics continues, so that technological gains better translate into practical realities. It lies in the shared responsibility of being transparent in AI. By investing in innovation, fostering collaboration, and demonstrating commitment towards ethical practices, global communities can ensure accountability, equity, and alignment with human best interests for AI technologies. After all, this is a moment to act, because what is decided today will determine the trajectory of the use of AI for generations to come.

REFERENCES

1. Ville Aula and Tero Erkkilä. Ai and transparency. In *Handbook on Public Policy and Artificial Intelligence*, pages 170–180. Edward Elgar Publishing, 2024.
2. Madalina Busuioc. Accountable artificial intelligence: Holding algorithms to account. *Public Administration Review*, 81(5):825–836, 2021.
3. Giovanni Sartor, Francesca Lagioia, et al. The impact of the general data protection regulation (GDPR) on artificial intelligence. 2020. https://doi.org/10.2861/293
4. John Frank Weaver. Ai issues raised by the California Consumer Privacy Act. *The Journal of Robotics, Artificial Intelligence & Law*, 3, 2020.
5. Anne L. Washington. How to argue with an algorithm: Lessons from the compas-propublica debate. *Colo. Tech. LJ*, 17:131, 2018.
6. Kimberly A. Houser and W. Gregory Voss. GDPR: The end of Google and Facebook or a new paradigm in data privacy? *Rich. JL Tech.*, 25:1, 2018.
7. Nagadivya Balasubramaniam, Marjo Kauppinen, Kari Hiekkanen, and Sari Kujala. Transparency and explainability of ai systems: ethical guidelines in practice. In *International Working Conference on Requirements Engineering: Foundation for Software Quality*, pages 3–18. Springer, 2022.
8. Mark Walker. Consequentialism, deontology, and artificial intelligence safety. In *Artificial Intelligence Safety and Security*, pages 411–421. Chapman and Hall/CRC, 2018.
9. Patricia Gomes R^ego de Almeida, Carlos Denner dos Santos, and Josivania Silva Farias. Artificial intelligence regulation: A framework for governance. *Ethics and Information Technology*, 23(3):505–525, 2021.
10. Thomas Søbirk Petersen. Ethical guidelines for the use of artificial intelligence and the challenges from value conflicts. *Etikk i Praksis-Nordic Journal of Applied Ethics*, (1):25–40, 2021.
11. Vikas Hassija, Vinay Chamola, Atmesh Mahapatra, Abhinandan Singal, Divyansh Goel, Kaizhu Huang, Simone Scardapane, Indro Spinelli, Mufti Mahmud, and Amir Hussain. Interpreting black-box models: A review on explainable artificial intelligence. *Cognitive Computation*, 16(1):45–74, 2024.
12. William G. Baxt. Application of artificial neural networks to clinical medicine. *The Lancet*, 346(8983):1135–1138, 1995.
13. Jeremy Petch, Shuang Di, and Walter Nelson. Opening the black box: The promise and limitations of explainable machine learning in cardiology. *Canadian Journal of Cardiology*, 38(2):204–213, 2022.
14. Liat Antwarg, Ronnie Mindlin Miller, Bracha Shapira, and Lior Rokach. Explaining anomalies detected by autoencoders using shapley additive explanations. *Expert Systems with Applications*, 186:115736, 2021.
15. Umut Turksen, Vladlena Benson, and Bogdan Adamyk. Legal implications of automated suspicious transaction monitoring: enhancing integrity of AI. *Journal of Banking Regulation*, 1–19, 2024.
16. Josh Meltzer and Aaron Tielemans. *The European Union AI Act*. Bruselj: Brookings Institution, 2022.
17. Frederik Zuiderveen Borgesius et al. *Discrimination, Artificial Intelligence, and Algorithmic Decision-Making*. Council of Europe, Directorate General of Democracy, page 42, 2018.
18. Ben Shneiderman. Bridging the gap between ethics and practice: Guidelines for reliable, safe, and trustworthy human-centered ai systems. *ACM Transactions on Interactive Intelligent Systems (TiiS)*, 10(4):1–31, 2020.

19. Heike Felzmann, Eduard Fosch-Villaronga, Christoph Lutz, and Aurelia Tam'o-Larrieux. Towards transparency by design for artificial intelligence. *Science and Engineering Ethics*, 26(6):3333–3361, 2020.
20. Karen Yeung. Recommendation of the council on artificial intelligence (OECD). *International Legal Materials*, 59(1):27–34, 2020.
21. Francesc Pedro, Miguel Subosa, Axel Rivas, and Paula Valverde. Artificial intelligence in education: Challenges and opportunities for sustainable development. *UNESCO Working Papers on Education Policy*, 7: 1–48, 2019.

Index

A

aberrant patterns, 21
accountability, 4, 8, 33, 34, 37, 174, 179, 180, 192, 197, 204
accuracy, 3, 4, 8, 16, 22, 24, 29, 30
admissibility, 11, 22, 23, 24, 26
adversarial, 12, 24, 25
AI forensic tool, 85
AI-Powered Systems, 162, 163
algorithmic bias, 194, 201, 202
AmCache, 36
Anaphora resolution, 54, 55
anomaly detection, 3, 34, 36, 98, 113, 116
anonymization, 73
APK Analysis, 151, 154, 156
application logs, 36
Artificial Intelligence, 126, 135, 173–175, 191, 192, 193
Association of Chief Police Officers, 43
ASTM E1188–11, 43
authentication, 52
automated analysis, 34
automated data analysis, 3
automatic text processing, 50
automating data collection, 83
automation, 82, 119
Autonomic Computing Initiative (ACI), 77
autonomic manager, 77
autonomous computer system, 77
autonomous computing, 78, 80
autopsy, 7

B

behavioral analytics, 6
behavioral forensics, 126, 132
behavioral profiling, 117
bias, 174, 175, 179
big data, 16, 97
black box, 198, 201, 208
blockchain, 164
blockchain technology, 5, 88
Budapest Convention, 23

C

California Consumer Privacy Act (CCPA), 193, 194, 208
case management, 160, 162, 166
Cellebrite Guardian, 91, 93
centralization of data, 89
chain of custody, 22, 24
chatbot, 72–79
classification analysis, 112
client portals, 164
cloud computing, 89
cloud services, 97
clustering analysis, 113
clustering model, 60
cognitive science, 77
Colloquium for Information Systems Security Education, 39
COMPAS, 15, 21
comprehensive analysis, 83
computational challenges, 121
computational resources, 11, 19
computational technologies, 52
computer resource management, 75
CoNLL, 63
contextual understanding, 50, 52
continuous professional development, 30
conversational artificial intelligence (CAI), 75, 76, 77
convolutional neural networks, 36
corruption, 135, 136, 137, 139
cost-effectiveness, 93
countering cybercrime, 83
credibility, 22, 24
crime scene evidence, 83
criminal evidence collection, 96
criminal justice system, 21, 31
critical evidence, 24
cryptographic technique, 89
cryptography, 87
Cuckoo Sandbox, 156
culture, 130, 131
cyberattacks, 10, 16
cyberbullying, 14
cybercrime, 96, 115, 116, 168
cybercriminal, 12, 25
cybersecurity, 96, 116
cyber threat intelligence, 6

D

data acquisition, 36, 40
data analysis, 96, 115
data analytics, 163, 164
data center management, 77
data collection, 89
data extraction, 34, 50

data management planning, 34
data mining, 132
data parsing, 50
data privacy, 174, 176, 181, 193, 194, 206
data quality, 11, 12, 13, 28, 121
data visualization, 6, 7, 104, 105, 107, 111
database forensics, 89
Daubert v. Merrell Dow Pharmaceuticals case, 23
DBpedia, 59, 61
DDoS attack, 17, 59, 60
debiasing methods, 73
decentralized ledger technology, 88
decision trees, 103, 110
deep belief networks, 36
deep learning, 173
Defense Production Act, 86
department of homeland security, 87
digital communications, 22
digital divide, 167
digital evidence validation, 118
digital forensic investigation, 83
digital forensics, 96, 115, 173, 174, 175, 192
digital open-source information, 42
dimensionality reduction, 107, 113, 119
document automation, 164
document corpus, 128, 123
DPDP Act, 90
dynamic analysis, 151, 154, 156

E

E-Courts Project, 165
E-Filing Systems, 164, 165
ELIZA, 72, 74
EnCase, 7, 91
encryption, 39
enforcement agencies, 92
escalation management, 76
ethical challenges, 122, 174, 180, 181
ethical frameworks, 192, 196, 207
EU AI Act, 194, 203, 206
European Union, 90
evidence analysis, 173, 180, 181
expert system, 84
explainability, 15, 24, 29
Explainable AI (XAI), 12, 33, 75, 76, 79, 80, 199, 204, 207

F

fairness, 135, 136, 192, 196, 201
feature engineering, 121, 122
feedback-based control, 77
financial transactions, 22, 24
flair model, 56, 57, 62, 63, 66
forensic image, 85
forensic intelligence, 126, 136
fraud detection, 96, 121
FTK, 91

G

General Data Protection Regulation (GDPR), 14, 21, 29, 73, 193, 194, 203
generative adversarial networks, 60, 61
generative models, 134
Genymotion, 153, 155
geopolitical entity, 55
government behavior, 129
government control, 86

H

habits, 127, 131
heterogeneity, 34
heterogeneous forensic tools, 39
hiding evidence, 83
high-risk AI system, 88
historical data, 126, 134
honesty, 135, 136
honey trapping, 84
HTML, 50
human autonomy, 197, 207
human oversight, 175, 179, 180
human-readable standard, 34

I

identification, 9, 13, 17, 21, 27, 28, 29, 51, 52, 57, 58, 59, 62, 66
IEEE Global Initiative, 29
image analyzer, 178
image and video analysis, 3, 5
image recognition, 27, 30
imaging and video forensics, 116
incident response, 116
inconclusive evidence, 180
indicators of compromise, 149, 151
individual behavior, 127
information space, 127, 130
intellectual property, 197, 199, 208
intent management, 76
intent prediction, 76, 77
internet of everything, 78
Internet of Things (IoT), 9, 17, 90
interoperability, 34, 43
Interpol, 56
intruders, 83
Intrusion Detection Systems (IDS), 116
intrusion detection, 84
IoT and Wearable Devices, 97
IPR protection, 88
ISO/IEC, 43

J

jurisdictional differences, 23

K

K-Means Clustering, 100, 113
K-nearest neighbor, 85
knowledge recognition, 84

L

law enforcement, 162
lawyers, 160
learning phase, 85, 86
legal research, 159, 163
legal standards, 19, 22, 31
legal support staff, 161, 162
lemmatization, 52
linear regression, 101, 102
litigants, 161
logistic regression, 102
LSTMs, 60

M

machine learning algorithms, 2–5, 7, 96, 98, 115, 174
malicious activities, 12, 17, 24, 25
malware, 149, 150
malware analysis, 3, 115, 150, 151
malware detection, 3, 83
manual analysis, 2
MAPE-K loop, 80
memetic algorithms, 36
memory forensics, 85, 86
metadata, 84
methodology, 131, 132
Microsoft Corporation, 55
Microsoft Ireland case, 23
misguided evidence, 180, 181
MIT Media Lab, 21
Mitsuku, 74
mobile devices, 97
model card, 44, 46
morphological analysis, 52, 53
multidimensional evolution, 34
multidimensional feature, 60

N

Named entity recognition (NER), 51, 54, 55, 56, 62, 63, 64, 65, 66
National Institute of Standards and Technology (NIST), 42, 43, 87
national science foundation, 87
national security memorandum, 87
national strategy for artificial intelligence, 88
natural language generation (NLG), 49, 75
Natural Language Processing (NLP), 2, 5, 12, 27, 49, 50, 51, 52, 55, 56, 61, 62, 63, 65, 72, 74, 75, 76, 82
natural language understanding (NLU), 49, 74, 77
network logs, 36, 97
network traffic analysis, 3, 83
NITI Aayog, 88
noisy data, 13
non-AI forensic tool, 84
non-transparent complex algorithm, 23

O

online dispute resolution, 164
ontology, 51, 57, 58, 61, 62, 65, 66
organizational behavior, 128
OSINT, 6
overfitting and underfitting, 101

P

P2P networks, 77
Paliscope, 7
pattern recognition, 3, 7, 34, 37, 84, 93
pattern validation, 50
permissions analysis, 154, 157
placement module, 78
Ponzi scheme, 22
POS-tagging, 52
predictive analysis, 82, 118, 164, 174
predictive policing, 21, 22, 29
presentation, 9, 10, 22
preservation, 9
price discovery mechanism, 88
Principal Component Analysis (PCA), 107, 119, 122
privacy, 12, 14, 19, 20, 21, 23, 28, 29, 30, 31
privacy concerns, 4
privacy-preserving technique, 87, 89
probabilistic reporting, 175
process monitoring, 84
proof-of-work (PoW), 79
PyTorch, 5

Q

QoS, 75
quantum computing, 75, 77, 79

R

RAM, 83
random forest, 59, 61, 85
ransomware, 150
RDF Query Language, 58, 59, 62

real-time processing, 16, 17
recurrent neural networks (RNNs), 36, 60
regular expressions (Regex), 50
regulatory compliance, 88
Reinforcement Learning (RL), 100
reliability, 12, 18, 22, 24, 26
replacement and formatting, 50
reproducibility, 24
research coordination network, 87
resource description framework (RDF), 63
response generation, 76
reverse engineering, 151, 155
robust optimization-based approaches, 25
robustness, 50, 59
ROC Curve Analysis, 175, 176, 177

S

sandbox, 151, 156
santoku, 153, 154, 155
scalability, 3, 98, 122
scientific evidence, 23
scientific working group on digital evidence, 43
sectoral regulatory framework, 88
security logs, 36, 43
self-adaptive system, 77
self-awareness, 78
self-configuration, 77, 78
self-improving system, 77
self-management, 77
semantic analysis, 52, 54, 55, 65
Semantic Role Labeling (SRL), 54, 55
semantic web technologies, 50, 51, 57, 58, 61, 62, 63, 65, 66
sentiment analysis, 76, 77
service-oriented architecture (SOA), 79
ShimCache, 36
SleuthKit, 91
smart acquisition work, 85
smart image, 85
social actors, 133, 139
social behavior, 126, 130, 132
social media platforms, 97
social media, 130
social network, 133, 139
socioeconomic, 15, 21, 29
sophisticated analytics, 76
SoTA, 72
SpaCy, 52, 53, 54, 55, 56, 57
SPARQL, 58, 59
speech recognition, 74
speech-to-text (STT), 74
spyware, 150
SQL injection, 59
stacked auto-encoders, 36
stakeholder trust, 191, 201, 204
static analysis, 151, 154, 157
stemming, 52
supervised learning, 99
support vector machine (SVM), 59, 61, 103, 115
swarm intelligence, 118
syntactic analysis, 52, 53, 54
system logs, 36
System Resource Usage Monitor (SRUM), 36
systemic complexity, 34, 35, 46

T

TensorFlow, 5, 7
testing hypothesis, 86
text analysis, 74
text-to-speech (TTS), 74
textual evidence analysis, 117
time series, 60, 61
time series analysis, 114
timeline reconstruction, 116
traceability, 88
trackable logs, 91
transmission control protocol, 43
transnational AI-based investigation, 23
transparency, 2, 4, 8, 11, 15, 22, 24, 28, 29, 30, 31, 174, 175, 178, 191, 192, 193
trojan, 150
t-SNE, 107, 123

U

ubiquity, 33, 34
UFDR data, 91
Uniform Manifold Approximation and Projection (UMAP), 108, 123
Universal Forensic Extraction Device (UFED), 7
unsupervised learning, 100, 113
URL, 50

V

validation spans, 24
verification, 84
veritone redact works, 92
virtual courtrooms, 164, 165
virtual legal assistants, 163, 166
virus, 150
VirusTotal, 151, 156
volatile data, 83
volatile memory, 84, 85
volatility, 7, 39
volatility memory analyzer, 85
VolMemLyzer, 85
voyeuristic material, 90
vulnerabilities, 87

W

web-based platform, 92
Web Ontology Language (OWL), 58, 59
windows artifacts, 36
word frequency, 131
world wide web consortium, 58
worm, 150

X

X-Ways, 7

For Product Safety Concerns and Information please contact our EU representative GPSR@taylorandfrancis.com
Taylor & Francis Verlag GmbH, Kaufingerstraße 24, 80331 München, Germany

www.ingramcontent.com/pod-product-compliance
Ingram Content Group UK Ltd.
Pitfield, Milton Keynes, MK11 3LW, UK
UKHW020200200926
475702UK00004B/541

* 9 7 8 1 0 3 2 8 4 4 6 8 8 *